```
541.22 Rod
Rodger.
Molecular geometry.
```

**The Lorette Wilmot Library
Nazareth College of Rochester**

Molecular Geometry

*I have seen the task which God has given
men to occupy themselves.
He has made everything beautiful in its time.
He has also set eternity in their hearts;
yet they cannot understand what God
has done from beginning to end.
I know there nothing better for them than
to rejoice and do good while they live.
I know that everything God does will remain forever;
there is nothing to add to it
and there is nothing to take from it.
God does it so men will revere Him.*

Ecclesiastes 3

Molecular Geometry

Alison Rodger

Mark Rodger

Butterworth-Heinemann Ltd
Linacre House, Jordan Hill, Oxford OX2 8DP

⌥ A member of the Reed Elsevier plc group

OXFORD LONDON BOSTON

MUNICH NEW DELHI SINGAPORE SYDNEY

TOKYO TORONTO WELLINGTON

First published 1995

© Butterworth-Heinemann Ltd 1995

All rights reserved. No part of this publication
may be reproduced in any material form (including
photocopying or storing in any medium by electronic
means and whether or not transiently or incidentally
to some other use of this publication) without the
written permission of the copyright holder except
in accordance with the provisions of the Copyright,
Designs and Patents Act 1988 or under the terms of a
licence issued by the Copyright Licensing Agency Ltd,
90 Tottenham Court Road, London, England W1P 9HE.
Applications for the copyright holder's written permission
to reproduce any part of this publication should be addressed
to the publishers

British Library Cataloguing in Publication Data
A catalogue record for this book is available from the British Library

ISBN 0 7506 2295 4

Library of Congress Cataloguing in Publication Data
A catalogue record for this book is available from the Library of Congress

Printed in Great Britain by Clays Ltd, St Ives plc

CONTENTS

Preface		*ix*
Periodic Table		*xi*

1	Definition and Determination of Molecular Geometry	1
	Introduction	1
	1.1 What is molecular geometry?	2
	1.1.1 Isomerism	6
	1.2 Factors determining molecular geometry	8
	1.3 Theoretical models	14
	1.3.1 Molecular orbital and valence bond theories	14
	1.3.2 Steric-plus-electronic methods	22

2	A Unified View of Stereochemistry and Stereochemical Changes	35
	Introduction	35
	2.1 Point symmetry	36
	2.1.1 Point symmetry operations	36
	2.1.2 Point symmetry groups - formalism	37
	2.1.3 Chiral and achiral point groups	39
	2.1.4 Examples of point symmetries	41
	2.2 Determination of symmetry adapted functions	41
	2.2.1 Molecular orbitals and molecular orbital energy level diagrams from symmetry	41
	2.2.2 Vibrations	54
	2.2.3 Symmetries of wavefunctions	56
	2.3 ML_n geometries and their interconversion	57
	2.3.1 ML_n symmetry, geometry and stability	57
	2.3.2 Stereochemical changes	60
	2.3.3 The classical symmetry selection rule procedure	63

3	The Geometry of Molecules of Second Row Atoms		71
	Introduction		71
	3.1	Geometries of ML_n, $n=2,3,4$	74
	3.2	Carbon based chemistry	77
		3.2.1 Geometry about a C	77
		3.2.2 Stereoelectronic effects	80
	3.3	Boranes	85
		3.3.1 Bonding schemes for boranes	86
4	Main Group Elements Beyond the Second Row		95
	Introduction		95
	4.1	Halogen compounds	97
	4.2	The middle of the p block	100
	4.3	The left hand side	102
5	Complexes of Transition Metals and f-block Elements		105
	Introduction		106
	5.1	Transition metal complexes	107
		5.1.1 A survey of transition metal complexes by coordination number	107
		5.1.2 Determining transition metal complex geometries: an overview	111
		5.1.3 Crystal field theory (CFT)	112
		5.1.4 Ligand field theory (LFT)	116
		5.1.5 Steric versus electronic effects on transition metal complex geometry	120
	5.2	Lanthanides and actinides	132
6	Organometallic Compounds and Tranistion Metal Clusters		139
	Introduction		139
	6.1	Metal carbonyls	140
	6.2	Transition metal clusters	142
		6.2.1 The metal polyhedron	143
		6.2.2 Metal polyhedron plus ligand polyhedron	149
	6.3	Some examples	155

7	Macromolecules: Small Changes and Large Effects	163
	Introduction	163
	7.1 DNA	164
	7.2 Proteins	171
	7.3 The final word	174

Appendix 1	Rules for Multiplication of Point Symmetry Operations	177
Appendix 2	Generating Point Groups	179
Appendix 3	The Jahn-Teller Theorem	183

Index   185

PREFACE

Chemistry is a subject which manages to be, at the same time, both fascinating and frustrating. There appears to be a very elegant structure underlying the diverse range of chemical phenomena that is found, yet this structure seems to be well and truly hidden by the sheer complexity of the behaviour observed: why do molecules react one way under some conditions, but give completely different products under other conditions? why do some atoms adopt a range of different oxidation states and geometries, while others always behave in the same way? To the uninitiated, more experienced chemists can sometimes just add to the frustration. Their "chemical intuition" - a convenient euphemism for years of experience - often enables them to deduce the correct geometry for a molecule, or product for a reaction, or result from an experiment, with no apparent basis for the conclusion. We have tried to capture some of the rationale that guides chemists in reaching their conclusions, and put onto paper some of that subconscious filing system that is necessarily developed over years of acquiring data about chemistry.

Such an aim is clearly ambitious; perhaps even overly ambitious. Never-theless, it is our hope that we have been able to present some guidelines that will help the reader in perceiving the underlying structure of chemistry. We have tried to show that molecular geometry can be understood in terms of a few underlying principles that are always operative, but whose relative importance varies. As a result, the different theories of molecular geometry can be seen not to be in opposition to each other; rather, they concentrate on systems and conditions under which different subsets of these governing principles dominate.

The main focus of this book is on the arrangement of atoms about a single atom, though the final part of Chapter 3 and almost the whole of Chapter 6 are devoted to molecules in which the central atom is replaced by a cluster of atoms. Also, in Chapter 7 some of the consequences of small, local variations in geometry are pursued for the macromolecules DNA and proteins. The language used throughout the book is established in Chapters 1 and 2. Some readers may find Chapter 2 to be too formalistic. It is devoted to the symmetry arguments that underpin a great deal of "chemical intuition", and we feel it is important to show that such arguments are rigorous, and to give examples of how they can be quantified. Many readers may prefer to use this chapter on a "need-to-know" basis, skimming through at first and referring back to it only when and if they need the detail for later chapters. In this way we hope the book will be useful to chemists at different stages of development by being able to be read at different levels. We have assumed an understanding of the concepts covered in first year chemistry courses. The emphasis has been biased in favour of parts of chemistry that we ourselves found (and find) most difficult to understand.

Finally, "Molecular Geometry" owes its existence to our parents who set us on the path of learning; to those who taught us most patiently, especially Eddie Schipper and Bob Gilbert; to Brian Johnson who believed we could do it; to the Chemistry Department at Stanford University who saw the beginning and the end of the project; and to John Freeman who turned some preliminary diagrams into illustrations. Our thanks go to them.

Alison Rodger   Mark Rodger

August, 1994

PERIODIC TABLE OF THE ELEMENTS

1 H 1.008																	2 He 4.003
3 Li 6.94	4 Be 9.01											5 B 10.81	6 C 12.01	7 N 14.01	8 O 16.00	9 F 19.00	10 Ne 20.18
11 Na 22.99	12 Mg 24.31											13 Al 26.98	14 Si 28.09	15 P 30.97	16 S 32.06	17 Cl 35.45	18 Ar 39.95
19 K 39.10	20 Ca 40.08	21 Sc 44.96	22 Ti 47.90	23 V 50.94	24 Cr 52.01	25 Mn 54.94	26 Fe 55.85	27 Co 58.93	28 Ni 58.71	29 Cu 63.54	30 Zn 65.37	31 Ga 69.72	32 Ge 72.59	33 As 74.92	34 Se 78.96	35 Br 79.91	36 Kr 83.80
37 Rb 85.47	38 Sr 87.62	39 Y 88.91	40 Zr 91.22	41 Nb 92.91	42 Mo 95.94	43 Tc (97.91)	44 Ru 101.07	45 Rh 102.91	46 Pd 106.4	47 Ag 107.87	48 Cd 112.40	49 In 114.82	50 Sn 118.69	51 Sb 121.75	52 Te 127.60	53 I 126.90	54 Xe 131.30
55 Cs 132.91	56 Ba 137.34	57-71 *Ln	72 Hf 178.49	73 Ta 180.95	74 W 183.85	75 Re 186.2	76 Os 190.2	77 Ir 192.2	78 Pt 195.09	79 Au 196.97	80 Hg 200.59	81 Tl 204.37	82 Pb 207.19	83 Bi 208.98	84 Po (208.98)	85 At (209.99)	86 Rn (222.02)
87 Fr (223.02)	88 Ra (226.03)	89-103 *An	104 Rf (261.11)	105 Ha (262.11)	106 Sg (263.12)	107 Ns (262.12)	108 Hs (265.13)	109 Mt (266.14)	110	111	112						

*Ln for Lanthanides

57 La 138.91	58 Ce 140.12	59 Pr 140.91	60 Nd 144.24	61 Pm (144.91)	62 Sm 150.35	63 Eu 151.96	64 Gd 157.25	65 Tb 158.92	66 Dy 162.50	67 Ho 164.93	68 Er 167.26	69 Tm 168.93	70 Yb 173.04	71 Lu 174.97

*An for Actinides

| 89
Ac
(227.03) | 90
Th
232.04 | 91
Pa
(231.04) | 92
U
238.03 | 93
Np
(237.05) | 94
Pu
(244.06) | 95
Am
(243.06) | 96
Cm
(247.07) | 97
Bk
(247.07) | 98
Cf<
(251.08) | 99
Es
(252.08) | 100
Fm
(257.10) | 101
Md
(258.10) | 102
No
(259.10) | 103
Lr
(262) |
|---|---|---|---|---|---|---|---|---|---|---|---|---|---|---|

Atomic weights calculated from data in *Table of Radioactive Isotopes* by E. Brown, R.B.Firestone and V. Shirley (Wiley 1986) and masses in *The 1983 Atomic Mass Evaluation* by A. H. Wapstra and G. Audi (Phys. Rev. A432, 1985,1). For the unstable elements the atomic mass of the most longlived isotope is given within parenthesis. For elements 103 up, IUPAC 1994 recommends the symbols Unq (104), Unp (105), Unh (106), Uns (107), Uno (108), Une (109), and Uun (110); we prefer the symbols shown above, and, in addition, Ln for the lanthanides (Z=57-71) and An for the actinides (Z=89-103).

CHAPTER 1

Definition and Determination of Molecular Geometry

Contents

Introduction		1
1.1	What is molecular geometry?	2
	1.1.1 Isomerism	6
1.2	Factors determining molecular geometry	8
1.3	Theoretical models	14
	1.3.1 Molecular orbital and valence bond theories	14
	Molecular orbital theory for H_2	15
	Valence bond theory for H_2	17
	Simple molecular orbital theory for second row homonuclear diatomics	17
	Simple molecular orbital theory for second row heteronuclear diatomics	21
	Concluding comments	21
	1.3.2 Steric-plus-electronic methods	22
	Determining coordination number	23
	Arrangement of ligands about the central atom	25
	Molecular mechanics	30

Introduction

The shape of a molecule may seem like one of its simplest properties, yet it can convey a wealth of information to any chemist who knows how to interpret it. This book is motivated by the belief that the first step towards understanding the chemistry of a molecule is to know its geometry and to begin to understand why it adopts that shape. Molecular shape arises as a balance between various steric and electronic-structural effects; both of these types of effect also constrain the possibilities for intermolecular interactions, reactivity and spectroscopy. We shall seek to examine the underlying principles that govern the shape or shapes adopted by any given molecule, and in so doing will assess the advantages and limitations of different approaches to molecular geometry.

Once we understand the principles that determine molecular geometry, we may begin to interpret the clues that molecular geometry provides and make predictions about new systems. It is not intended that this book should be a comprehensive case-by-case study. There are a number of excellent general texts that perform that role[1-6] and give a balanced view of the relative importance of any given geometry. By concentrating on principles rather than examples, we hope to provide a unified framework in which the extensive chemical literature can be read with understanding. As with arrow-pushing rationalisations of organic reaction pathways, we shall never be able to be correct every time, but a few exceptions are easier to remember than a book full of apparently unrelated facts, and the facts that do not fit into a neatly constructed framework tend to be the goad that leads to the next step in understanding. In this way we hope to address the complaints most frequently voiced by students studying inorganic chemistry, namely that there is so much to learn and apparently no logical reason why the relevant molecules behave as they do.

In the rest of this chapter we shall give an overview of molecular geometry, and we begin by considering exactly what we mean by "molecular shape"; although the term is intuitively obvious, a precise definition can be surprisingly elusive. This will be followed by a discussion of the factors that determine molecular geometry and a brief résumé of some existing approaches to understanding molecular geometry. Existing approaches tend to be disparate and to apply only in certain circumstances - we shall endeavour to unify the ideas they contain in the rest of the book.

1.1 What is Molecular Geometry?

Upon being confronted by the question "what is molecular geometry?" most chemists would start to describe simple geometrical shapes that they associate with various molecules or parts of molecules. At the beginning of a book on molecular geometry we must consider just what we mean by a statement such as "methane is tetrahedral and looks like Fig. 1.1, having four C-H bonds pointing to the vertices of an imaginary tetrahedron".

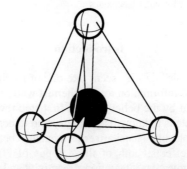

Fig. 1.1 Methane.

At best, such a picture is either the lowest energy arrangement of methane's atoms, or an average of the geometries methane really adopts. In reality methane is not so rigid, but has atoms that are constantly vibrating even in the solid state. In some instances, the average geometry is not an energy minimum. For example, the most stable geometry of "octahedral" d^9 metal complexes, e.g. $[Cu(H_2O)_6]^{2+}$ (Fig. 1.2), is tetragonal due to the Jahn-Teller effect (see §5.1.5). In other cases, a molecule may adopt different geometries depending upon its environment. For example, the drug Hoechst 33258 (Fig. 1.2) adopts a more planar structure when bound to DNA than when it is free. Furthermore, there are even situations in which what we refer to as the geometry of a molecule may depend on the experimental technique used to determine it. Many transition metal cluster compounds (see Chapter 6) have two or more geometries of very similar energy, and the dominant one may depend on whether we are investigating it in the solid or in solution, and what the temperature is. Further, the geometry of a particular molecule in a sample may change during the time scale of a measurement, so that what we observe is some average of the possibilities.[7,8] There are various ways to describe the geometry and symmetry of such non-rigid systems.[9,10] In this book, we shall pretend that molecules are more-or-less rigid with clearly identifiable geometries unless we explicitly confess to the contrary.

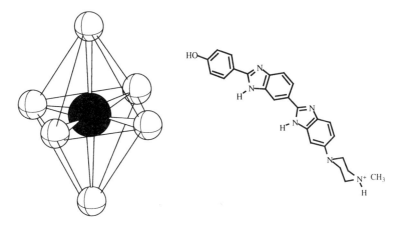

Fig. 1.2 Tetragonal $[Cu(OH_2)_6]^{2+}$ and Hoechst 33258.

Thus, we return to the question of what we mean by the geometry of a molecule. Are we identifying geometry with local minimum energies on the complicated potential energy surface that describes the energy of a collection of atoms as a function of atomic positions, or are we talking about some average geometry? And if it is the latter, then how are we performing the averaging (arithmetically, geometrically, root mean square ...)? We should also stop to ask what role the electrons play in all of this. So far we have described molecular geometry in terms of the location of the atomic nuclei, but it is the electrons that occupy the major part of a molecule's volume. Although some of these distinctions may seem to be insignificant, some important developments in our ability to predict chemical behaviour during the last decade have actually depended upon recognising the difference between these different descriptions of molecular shape or geometry.

Now, having acknowledged that we are asking a very complicated question, it is better to begin by ignoring most of these problems and consider idealised geometries that correspond to our unthinking answers to the question "What is the geometry of ...?". Instead of being apologetic about this simplification, let us extend the idea of idealised geometries; for example, utilising the similarities between highly symmetric molecules such as CH_4 and less symmetric ones such as chloroform, $CHCl_3$, and the amino acid glycine NH_2CH_2COOH (Fig. 1.3).

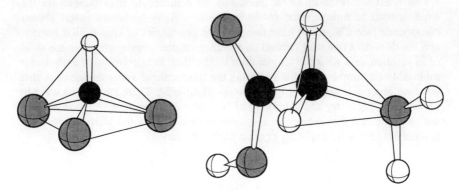

Fig. 1.3 Chloroform, $CHCl_3$, and the amino acid glycine NH_2CH_2COOH. C is black, H is white and other atoms are shaded.

In many situations it is convenient to describe the shape of a molecule by focusing on a subset of the atoms and associating an idealised high symmetry shape or *template* with these atoms; the selected atoms lie at or near to the vertices of this shape. The appropriate template for methane, chloroform and glycine would then be tetrahedral about the first C. Similarly, we might describe the sugar glucose in terms of a cyclohexane template (Fig 1.4); and *trans*-$[Co(NH_3)_4Cl_2]^+$ (Fig. 1.5) as octahedral. The usefulness of associating a molecule and a template will depend on the extent of distortion from the template and the property being discussed, but it enables us to begin to systematise the vast array of molecular shapes that arise in nature. We then add to the template picture the details of which atoms are bonded together, the length of the bonds, and the various bond and torsion angles.

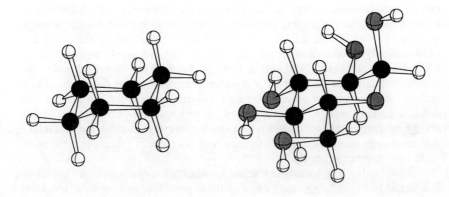

Fig. 1.4 Cyclohexane and glucose. Atom key as for Fig. 1.3.

Definition and Determination of Molecular Geometry

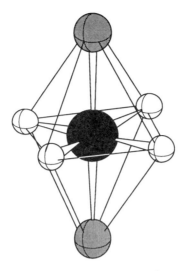

Fig. 1.5 *Trans*-[Co(NH$_3$)$_4$Cl$_2$]$^+$.

One further generalisation is possible when talking in terms of templates. If we take the template as a shape on which we build the molecule, then we can also construct a molecule by locating atoms on some, but not necessarily all of the vertices, with "holes" occupying the remaining vertices. In this sense a square planar complex may be associated with an octahedral template as well as a square. Such flexibility is valuable when we consider metal complexes such as nickel (II) cyanide that may be either square planar or square pyramidal (Fig. 1.6), both of which may be associated with an octahedral template; it also forms the basis of Chapter 2 where the relationships between different templates is explored, and is used in Chapter 6 when we discuss the geometry of transition metal cluster compounds.

In the following discussions it will be useful, if somewhat arbitrary, to draw a distinction between the terms *structure* and *geometry*. Molecular *structure* will be taken to refer to the precise arrangement of atoms (*i.e.* the location of the atomic nuclei) within the molecule; molecular *geometry* will be used rather more loosely.

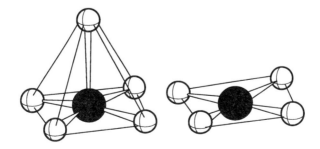

Fig. 1.6 Nickel (II) cyanide.

1.1.1 Isomerism

For small molecules, the molecular formula and a table of standard bond lengths and angles[11] may be all that is required to enable a fairly accurate estimate of the geometry. However, most molecular formulae have a number of possible different isomers that are consistent with the same bond lengths and angles. It is always possible (though not necessarily easy) to convert a molecule from one isomer or geometry to another, it simply requires energy. The types of isomerism possible are given below in order of increasing energy difference between isomers.

Optical Isomerism: Some molecules are not superposable[†] on their mirror images. This is the most subtle form of isomerism, and requires that the molecule has no centre of inversion, reflection plane or improper rotation axis (see §2.1). Such molecules are known as chiral (from the Greek word χειρ meaning hand). Two optical isomers (also called enantiomers) can only be distinguished from one another in the presence of a chiral influence: another chiral molecule or a chiral environment such as that provided by circularly polarised light. The simplest chiral molecule is hydrogen peroxide (Fig. 1.7), though the energy barrier between the two enantiomers is so small that all experiments detect both isomers equally, and no net chirality. The concept of chirality is crucial for biological systems. For example, the teratogenic properties of thalidomide (Fig. 1.7) are associated with only one of the enantiomers. Sugars (*e.g.* glucose, Fig. 1.4) and amino acids (Fig. 1.7), but excluding glycine (Fig. 1.3), are among the simplest common chiral molecules, and are component parts of some of the most complicated chiral molecules, including DNA and proteins (Chapter 7).

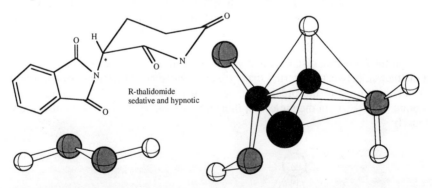

Fig. 1.7 H_2O_2, thalidomide (the S enantiomer, which is not illustrated, is the teratogen), and an amino acid. The amino acid side chain is depicted by the large black sphere.

Conformational Isomerism: Ethane, C_2H_6, requires about $4\,\text{kJ}\,\text{mol}^{-1}$ of energy for one CH_3 group to rotate with respect to the other about the C-C bond. The lowest energy *conformer* is the staggered one where the H's make the vertices of a

---

[†] Superposable is the correct term. Superimposable, which is more widely used, is misleading since non-identical items may be superimposed to facilitate comparison.

hexagon if the molecule is viewed along the C-C axis and squashed flat (more precisely we talk about projecting onto a plane perpendicular to the C-C bond) (Fig. 1.8). The other special conformation is the eclipsed form, in which the H's at the front obscure those at the back so that it projects onto a triangular shape; this is the highest energy possibility. If one continued twisting around the central C-C bond one would find three separate staggered conformations. In ethane these are all equivalent, but in a molecule such as 1,2-dichloroethane, with a chlorine attached to each C, they give rise to two chemically different staggered forms. We identify the two distinct forms as separate isomers - *gauche* (with chlorines at, *e.g.*, positions 1 and either 4 or 5 in the staggered ethane of Fig. 1.7[†]) and *trans* (with chlorines at, *e.g.*, positions 1 and 6). For 1,2-dichlorethane the torsional barrier is similar to that for ethane, ~ 4 kJ mol^{-1}; the proportion of molecules having sufficient energy to surmount this barrier will be approximately e$^{-(E/RT)}$, which amounts to about 50% at room temperature. In other words, roughly one out of every two torsional vibrations will result in a successful barrier crossing, and conformational interconversion will take place on a sub-nanosecond timescale.

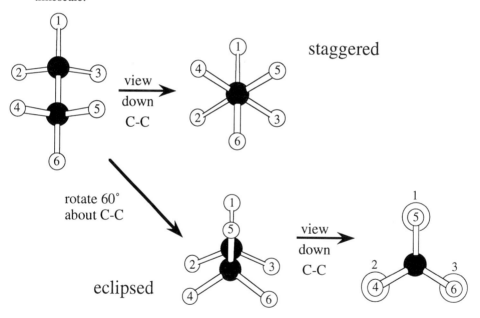

Fig. 1.8 Staggered and eclipsed forms of ethane.

Clearly, it is not possible to treat the conformers of 1,2-dichloroethane as separate chemical species under normal conditions, and so we may really only speak of stable conformers in this case, not of conformational isomers. It must be remembered, however, that this distinction is temperature dependent, and that if we were to work at liquid helium temperatures, then the stable conformers of 1,2-dichloroethane could be separated as conformational isomers. For

---

[†] The two *gauche* forms are, in fact, optical isomers of each another.

1,2-dichlorethene, chemically important conformational isomers, *cis* and *trans* (Fig. 1.9), may be identified since the energy barrier that prevents interconversion of the two forms is much higher than the typical energy associated with vibrational and torsional motions (~2.5 kJ mol^{-1} at room temperature). Thus interconversion of the two forms is slow and we can isolate them as distinct chemical species.

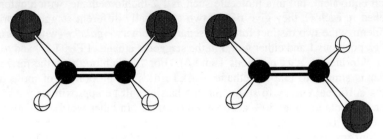

Fig. 1.9 *cis*- and *trans*-dichloroethene.

Structural Isomerism: The above types of isomerism all involve the same bond connections between atoms. However, C_3H_6 could exist as either cyclopropane or propene. Thus, C_3H_6 exhibits what is known as structural isomerism.

1.2 Factors Determining Molecular Geometry

Energy: Which molecular geometries are observed is determined solely by energetic considerations. One of the basic beliefs instilled into chemists is that the world around us is driven by two opposing trends: a tendency to minimise energy (enthalpy, H) and a tendency to maximise entropy (S). These two factors are combined into the concept of free energy, so that the driving force behind chemical and physical changes is then a "desire" to minimise the free energy; in most circumstances this is the Gibbs' free energy, G, defined as $G = H-TS$ where T is temperature. From a theoretical point of view it would be preferable to work with the internal energy (U) since this is the one thermodynamic energy that is a molecular property rather than a collective property and so in principle can be calculated directly. U is related to G *via* the definition $G = U-TS+PV$, where P is the pressure, and V is the volume of the system. It turns out that for a discussion of molecular geometry, the differences between U and G are not great, and one can get a long way by considering just the internal energy. This situation occurs primarily because molecular geometry is a property of an isolated molecule and the volume actually filled by the molecule is only a small part of the total volume available to it, so that changes in the space it fills due to conformational changes will cause very little difference to V. Again, for an isolated molecule the main contribution to the entropy associated with a given geometry arises from its vibrational motion; since we are neglecting vibrations in our discussion of geometry, it is also reasonable to neglect the TS term. Thus to a good approximation, stable molecular geometries may be associated with arrangements of atoms that will minimise the molecule's internal energy.

If we imagine all the possible isomers a molecule can adopt as being expressed in terms of a set of 3N-6 (or 3N-5 for linear molecules) Cartesian coordinates, then U may be expressed as a function of all those coordinates and pictured as a multidimensional potential energy (PE) surface. For diatomics this is the familiar plot illustrated in Fig. 1.10; for larger molecules it becomes impossible to visualise, but we can still calculate it and use it. Stable geometries are wells on the PE surface with walls substantially higher than RT, the energy readily available from thermal motion; transition states are saddle points; reactions or rearrangements occur when enough energy is provided for the molecule to climb out of the well and over a saddle point.

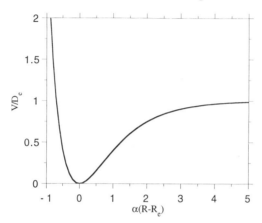

Fig. 1.10 Morse PE surface for a diatomic molecule. D_e = dissociation energy, $\alpha = \omega(\mu/(2D_e))^{1/2}$, ω = vibrational frequency, μ = reduced mass.

The question that now arises is how do we obtain this PE surface. Indeed, is it even a well-defined function? The discussion above has drawn heavily on classical thermodynamics for its justification, and so we must pause to ponder whether the considerations of quantum mechanics that govern vibrations and reactions in molecules might modify this picture. It turns out that we are justified in using the idea of a potential energy surface to identify stable conformers and their structure, and the reason we can do this is the Born-Oppenheimer Approximation (BOA). While this is not the place to treat the BOA rigorously, we should be aware of its existence and its significance for molecular geometry. The BOA follows from noting that electrons are so much lighter than nuclei that they will be able to respond essentially instantaneously to changes in the positions of the nuclei. This in turn means that it is possible to contemplate solving Schrödinger's equation (and hence finding out everything we need to know about the molecule) at each possible arrangement of the nuclei. We would then come up with an energy that characterises each configuration of the nuclei. The PE surface is just the collection of these energies.

The importance of the preceding discussion is that it provides us with a means of asking *why* a molecule adopts a particular geometry, and a criterion for determining what the possible geometries are. If a particular conformation is stable, then any distortion of the molecule away from that conformation must lead to an increase in energy. Equivalently, we may seek to ask what it is that makes the observed geometry favourable, *i.e.* what are the interactions that will lead to a lower energy. Of course, while the principle is easy to state, the practice is usually more complex, and so many different models have been suggested to explain why a molecule adopts one geometry and not another. These range from being extremely simple (yet based on hand waving arguments and often theoretically unjustified) to extremely complex (theoretically rigorous, but often too incomprehensible to give much insight). The important thing to determine in a given application is how near to *the truth* the answer needs to be, and how comprehensible it needs to be. Unfortunately, these two criteria are usually in opposition.

Two simple ideas can often help us circumvent many of the complexities in these energetic arguments. One is symmetry (see below and Chapter 2). The other relates to bond energies: are the bonds associated with a given geometry strong enough to withstand bond vibrations? The latter question may often be answered from simple orbital-overlap arguments (see §1.3).

Symmetry: Experience often leads a chemist to be able to guess what the most stable arrangement of a given set of atoms is. We need to ask what underlies that experience. One of the key features is that molecules usually adopt symmetric geometries if possible, and one can often guess whether retaining a particular symmetry element is energetically unfeasible. A symmetry element of a molecule is a geometric transformation (rotation or reflection or a combination of them) that leaves all measurable properties of that molecule unchanged. So, for example, if $[MnCl_6]^{4-}$ has high symmetry, it is unlikely to be planar with hexagonal symmetry because either the Cl's will be crowded or the Mn-Cl bonds too long, but an octahedral geometry seems to be a "sensible" option. In fact it can be proved rigorously[12] that high symmetry geometries are either stable (at an energy minimum), or a transition state (an energy maximum) between symmetric conformations. Thus for example, planar BF_3 is stable, but planar NH_3 is the transition state for the inversion of ammonia.

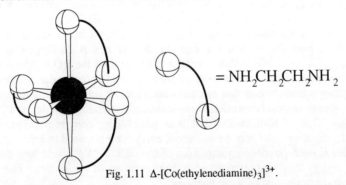

Fig. 1.11 Δ-$[Co(ethylenediamine)_3]^{3+}$.

We shall discuss symmetry much more fully in Chapter 2, but the key idea is that molecules are likely to adopt symmetrical shapes unless there is a good reason for them to distort away from their high symmetry template. Furthermore, when applying crystal structure data to isolated or solution phase molecules, it is justifiable to assume that the crystallographic environment may cause small distortions that would not persist in the isotropic environment provided by the solvent. For example, the crystal coordinates of [Co(ethylenediamine)$_3$]$^{3+}$ (Fig. 1.11)[13] indicate the molecule has no symmetry operations, but is very close to one having a three-fold rotation axis and three two-fold rotation axes perpendicular to it; one would expect these symmetry elements to reassert themselves in solution.

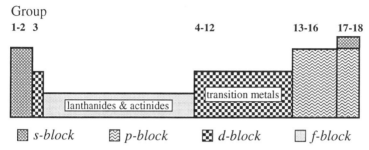

Fig. 1.12 Schematic periodic table indicating partially filled valence orbitals and the position of the lanthanides and actinides.

Orbital considerations and the periodic table: The other important key to our intuitive grasp of molecular geometry is the periodic table, which systematises a great deal of chemical data. The periodic table is arranged so that atoms with related chemical properties are lined up with one another either horizontally or vertically. The elements are numbered according to atomic number (number of protons and hence electrons in the neutral element). The columns, or groups, of the periodic table are then those atoms with the same type of electrons in the outer or valence shell (see below for definition). The columns are numbered 1 to 18, so the "*d*-block" is included in the numbering, but the "*f*-block" is excluded. If the lanthanides and actinides were as important to our lives as, *e.g.*, Fe, then the periodic table would we drawn as in Fig. 1.12 rather than the standard version shown inside the front cover.

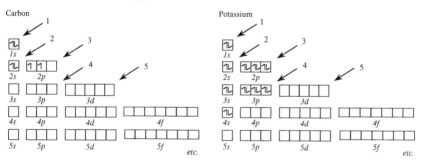

Fig. 1.13 Electron configurations for C and K. Arrows indicate order of orital occupancy.

Thankfully for the sanity of chemists, the distribution of the electrons of He, Li, … can, to a reasonable approximation, be described in terms of orbitals (functions describing the behaviour of a single electron) that look similar to the ones that can be determined exactly for hydrogen. So we may describe the electron configurations of *e.g.* carbon and potassium as in Fig. 1.13. The radial and angular behaviours of the orbitals of hydrogen are shown in Figs. 1.14 and 1.15 respectively. For H, the energy of an orital is dependent only on the principal quantum number n which correlates with the average distance from the nucleus of the electron.

The orbitals of other elements differ from those of hydrogen because electron-electron interactions must be included for all other elements. The most important difference between H orbitals and those of many-electron atoms, as far as the periodic table is concerned, is that the shape of an s orbital is such that an electron occupying it is, on average, closer to (penetrates further towards) the nucleus than a p electron of the same principal quantum number, so an s electron is more attracted to the nucleus and less protected (shielded) from it by other electrons than a p electron of the same principal quantum number. This means that for many-electron atoms the orbital energy depends not only on the principal quantum number, n, but also on the azimuthal quantum number, ℓ. p orbitals are therefore of slightly higher energy than the s orbitals of the same n, so the s orbitals are occupied first leading to the structure of the periodic table.

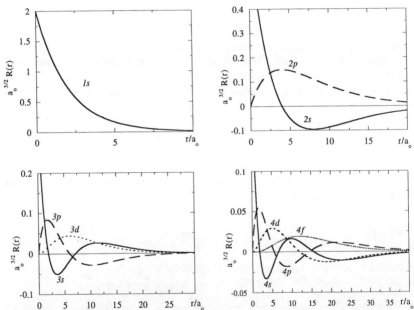

Fig. 1.14 Radial distribution $R(r)$[14] of H orbitals. r is the distance from the nucleus, a_o is the Bohr radius (5.292×10^{-11}m). Different scales are used for different principle quantum numbers, n. ℓ, the azimuthal quantum number, equals 0 for s orbitals, 1 for p orbitals, and 2 for d orbitals.

$$R(r) = -\sqrt{\left(\frac{2}{na_o}\right)^3 \frac{(n-\ell-1)!}{2n[(n+\ell)!]^3}} \left(\frac{2r}{na_o}\right)^\ell \exp\left(\frac{r}{na_o}\right) L_{n+1}^{2\ell+1}\left(\frac{2r}{na_o}\right) \text{ for}$$

$$L_\alpha^\beta(x) = (-1)^\alpha \frac{\alpha!}{(\alpha-\beta)!} \left[x^{\alpha-\beta} - \frac{\alpha(\alpha-\beta)}{1!} x^{\alpha-\beta-1} + \frac{\alpha(\alpha-1)(\alpha-\beta)(\alpha-\beta-1)}{2!} x^{\alpha-\beta-2} + ... \right]$$

The term "outer" or "valence" shell used above is in fact a loose label referring to all orbitals close in energy to the highest occupied and lowest unoccupied orbitals. For the second row of the periodic table this means the 2s and 2p orbitals. Further down the table it involves a mixture of principal quantum numbers. For example, the fourth row of the periodic table involves filling 4s orbitals, then 3d, then 4p. The 4s and 3d orbitals are very close in energy as is shown by the fact that it is usually the s rather than d electrons that are lost when e.g. Fe is ionised to Fe^{2+}.

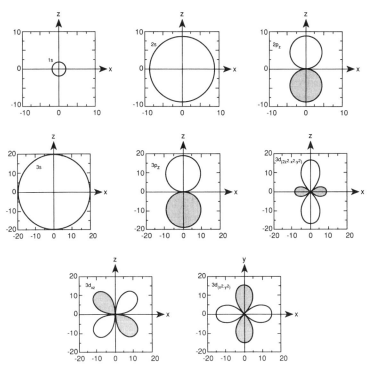

Fig. 1.15 "Shape" of *1s*, *2s*, *2p*, *3s*, *3p*, and *3d* hydrogenic atomic orbitals. The shapes plotted are derived as follows. The distance from the nucleus, r_{90}, within which 90% of the electron density of the occupied orbital is contained was determined. For each direction out from the nucleus, a vector whose length was proportional to the magnitude of the angular part of the orbital wavefunctions in the plane shown was drawn. The longest vector has length r_{90}. The heads of the vectors outline the shapes illustrated. $r_{90} = 2.7a_o$ for *1s*, $9.2a_o$ for *2s*, $9.1a_o$ for *2p*, $19.5a_o$ for *3s*, $18.4a_o$ for *3p*, and $15.8a_o$ for *3d*, where $a_o = 5.292 \times 10^{-11}$m. The angular functions are for np_z: $\cos\theta$; for $3d_{2z^2-x^2-y^2}$ $(3\cos^2\theta-1)/2$; for $3d_{xz}$: $\sin2\theta$; for $3d_{x^2-y^2}$: $\cos2\phi$, where θ is the angle from the vector to the *z* axis and ϕ is the angle to the *x* axis in the *x-y* plane. The diagrams for other *ao's*, e.g. $2p_y$ and $3d_{xy}$, are most simply determined by changing the axis labels on those illustrated. The shaded lobes have opposite sign functions from the unshaded ones. Contour plots have slightly different shapes.[14]

The similarity of chemical behaviour down a group results from the fact that most chemistry is determined by the valence electrons and atoms in the same group have the same type of valence electrons, albeit at greater distances from the nucleus as the principle quantum number increases. The differences down a group can often be related simply to this increase in size of the atom and consequent effects on electrons that are less tightly held (*i.e.* more polarisable). Another factor is the greater number of valence orbitals available.

1.3 Theoretical Models

Any discussion aimed at rationalising or predicting molecular geometries must begin with some degree of approximation or assumption. One approach that is conceptually helpful is to identify different "types" of energy, and to focus on their behaviour. For example, several models proceed by distinguishing between electronic and steric effects. The electronic effects relate to the sharing or donating of electrons and are normally associated with bonding, whereas the steric energy incorporates physical interactions normally associated with intermolecular interactions: the exclusion of other atoms from the volume occupied by a given functional group, the electrostatic interaction between polar functional groups *etc*. Many ways have been proposed for dividing the molecular energy into different categories, but we shall refer to all these models collectively as steric-plus-electronic models.

Although helpful, this division into "types" of interaction can be somewhat arbitrary, and so it is useful to consider such models in conjunction with molecular orbital (MO) and valence bond (VB) theories. These two theories give (in principle) rigorous quantum mechanical treatments of the system as a whole, and so avoid the problems introduced by arbitrary divisions. Unfortunately, by introducing such a level of rigour, one also tends to obscure physical insight, so that accurate MO and VB calculations are very difficult to interpret in simple geometric terms for polyatomic molecules. None-the-less, simple interpretations do arise for diatomic systems, and concepts that emerge from MO and VB treatments of diatomics prove to be very useful in developing, understanding, and justifying the various steric-plus-electronic models for more complicated polyatomic systems. MO and VB theories will therefore be considered in §1.3.1.

The purpose of the final part of this section is to present an overview of the more important non-quantum models of molecular geometry, and to indicate how they fit together in terms of the "types" of energy they consider or ignore. These models include valence shell electron pair repulsion theory, non-bonded radii, atom-atom repulsion, atom-atom interaction and molecular modelling approaches. The rest of this book will then illustrate how the particular application determines which energetic partition is appropriate, and hence which of the models would then be used.

1.3.1 Molecular Orbital and Valence Bond Theories

Although their own geometries appear trivial, diatomic molecules are invaluable for studying molecular geometry, since most of the basic bonding principles that hold more complicated molecules together can be deduced from them. This is a common trick in developing the theory of chemistry: simple systems that theoreticians can treat "properly" are used to develop the language with which we describe and understand much more complicated systems. Thus, the only atom for which we can solve the Schrödinger equation is the hydrogen atom, and, as mentioned above, we consistently use the hydrogen *ao's* as the basis for understanding the electronic structure of all atoms. In the same way, diatomic molecules are simple enough that we can hope to give a reasonably rigorous treatment of the molecular structure (albeit in terms of the *ao's* that came in turn from the H atom), and it is this treatment of the diatomic molecules that will provide us with the concepts and the language to describe much more complicated molecular systems.

The following rule of thumb is helpful in using the results from simpler systems for more complicated systems when quantum mechanics cannot be ignored (as is the case for electrons): where possibilities are alternatives, (is an electron in this or that state?) we add the respective wave functions, but when possibilities can happen at the same time (one electron is in this state while another is in that state) we multiply the states. These ideas are fundamental to the process of constructing the wavefunctions that describe the electrons in molecules.

What follows is a simplistic view of both MO and VB theory. In particular, we shall overlook many of the subtleties of electron spin except in so far as the Pauli principle allows only for double occupancy of each (spatial) orbital. Although spin-effects are important in quantitative calculations and in spectroscopic applications, they usually do not affect our qualitative understanding of molecular geometry. For a more complete treatment, see *e.g* references [14-18].

Molecular Orbital Theory for H_2

Simple MO theory describes the behaviour of each electron separately within its molecular environment by a function called a molecular orbital, or *mo*. (Note the use of lower case *mo* to refer to the orbital, but the upper case MO to refer to the theory as a whole.) The molecular environment of one electron includes the nuclei and an average of the field provided by the other electrons. Refinements of simple MO theory allow for correlated electron-electron interactions. We shall consider only simple MO theory. In general one expresses the *mo* for an electron in terms of the atomic orbitals (*ao's*) of the atoms and allows the electron to occupy any of the available *ao's*, though not necessarily with equal probability. The valence *ao's* are the ones most involved in any electron redistribution following molecule-formation, so we shall consider only these orbitals. The radial extent of the valence orbitals used for different atoms throughout this book were illustrated in Fig. 1.14.†

It is also necessary to consider the orientation in space of the orbital if we are to determine the overlap between orbitals. The magnitude of an *s* orbital is independent of direction. For *p*, *d*, *etc.*, however, we must consider the orbital shape. As chemists we are familiar with drawing *e.g.* dumbells for *p* orbitals. The precise meaning of these shapes is given in Fig. 1.15, but in general terms

---

† The electron density in an orbital is the square of the function illustrated, and the radial distribution function is the electron density in a spherical shell of radius r and thickness dr.

they depict the electron densities in different directions. The length of a line drawn from the origin to the curve is a measure of the electron density one would see looking out from the nucleus along that line. So, e.g, p_z has no electron density along the x axis and maximum density along z.

For the symmetric diatomic molecule H_2 each electron is equally likely to be either in the $1s$ orbital of H atom a or in the $1s$ orbital of H atom b. Thus, since these $ao's$ are alternatives (as discussed above) the $mo's$ in this basis set will be

$$\phi_s = [\psi(H_{1s}^a) + \psi(H_{1s}^b)] \quad \text{or} \quad \phi_s^* = [\psi(H_{1s}^a) - \psi(H_{1s}^b)] \quad (1.1)$$

ϕ_s has more electron density between the H's and ϕ_s^* has less. Thus, an electron in ϕ_s acts as negatively charged glue to hold the two positive protons together. Conversely, an electron in ϕ_s^* increases the proton-proton repulsion relative to two non-bonded H atoms, thus its occupancy is a driving force away from molecule formation. So ϕ_s is a bonding orbital and is lower in energy than ϕ_s^* which is antibonding. Thus we may draw a qualitative MO energy level diagram for H_2 as in Fig. 1.16. Once the $mo's$ have been determined, the electrons are then assigned to orbitals, with up to two (of opposite spin) in each. Now, there are two electrons in H_2 and we need to describe their simultaneous behaviour, so the lowest energy MO wavefunction for H_2 is given by the product of the electronic $mo's$:

$$\psi(H_2) = \phi_s(1)\phi_s(2) \quad (1.2)$$

The notation '(1)' in this equation means we are talking about the first electron being associated with this particular orbital. The second lowest energy function has one electron in $\phi_s^* = [\psi(H_{1s}^a) - \psi(H_{1s}^b)]$. The lowest energy arrangement of electrons is illustrated in Fig. 1.16.

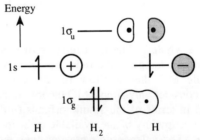

Fig. 1.16 H_2 valence MO energy level diagram illustrating, schematically, the electron density in each orbital (were it to be occupied); shaded parts have negative phase, unshaded parts have positive phase. Electron occupancy of ground state is indicated by arrows whose direction indicate spin.

As drawn Fig. 1.16 suggests that if both orbitals were filled, as would be the case for He_2, then the molecule would have the same energy as two separated atoms. In fact, the situation is slightly worse than that, as anti-bonding orbitals are always slightly more antibonding than bonding orbitals are bonding, so He_2 is actually less stable than 2He.

The above discussion of H_2 has used only the $1s$ orbitals on the H's. In fact, the presence of a second H atom distorts the $ao's$. This distortion could be described in terms of the electron being excited to spend part of its time in one of the higher energy *ao's*, so in a more accurate description we would include contributions from *2s, 2p ... ao's*. Of course, the higher in energy such orbitals are, the less likely such an excitation is to take place, and so the less significant will be the contribution to the ground state wavefunction. In general, the amount of interaction between orbitals on different atoms depends not only upon their spatial overlap, but also on the energy difference between them. Orbitals that are closest in energy interact most. Since the energy gap between *1s* and *2s* is large (see Fig. 1.19), the *2s* would have only a small effect on the lowest orbitals of H_2 and the relatively simple *mo's* given above provide a fairly accurate description. The situation is not so simple for the second row systems Li_2 and Be_2 as discussed below.

It should be noted that the energy of the molecule is not simply the sum of the orbital energies times their occupancy, since this ignores the interactions between electrons. The two energy terms that account for electron-electron interaction are the Coulomb energy (a positive destabilising energy resulting from the electrostatic repulsion of two negatively charged particles) and the exchange energy (a negative term resulting from the fact that electrons are indistinguishable from each other; it may be thought of as the quantum mechanical analogue of entropy, since it comes from recognising that rearranging the system by interchanging electrons leads to arrangements that are identical). These terms become important for transition metal systems (Chapter 5), where the energy gap between different *d* levels is small and the balance between coulomb and exchange energies may lead to the electrons singly occupying higher levels rather than doubly occupying the lowest energy levels in accord with the aufbau principle.

Valence Bond Theory for H_2

Whereas MO theory starts by describing single electrons, VB theory begins with pairs of electrons. It is motivated, at least in part, by the idea that a bond results from sharing a pair of electrons. The simple VB wavefunction for H_2 is, therefore, just the average of the two alternatives with one electron on each atom at any one time:

$$[\psi(H_{1s}^a)(1) \; \psi(H_{1s}^b)(2) \; + \; \psi(H_{1s}^a)(2) \; \psi(H_{1s}^b)(1)] \tag{1.3}$$

The advantage of this function over the MO one of Eq. 1.2 is that its dissociation limit is two H atoms. Its disadvantage is that it allows for no ionic electron distribution with more electrons on, say, H^a. An ionic term is therefore often added. In recent years VB theory has been developed to the point where accurate calculations can be performed.[19] For some systems, VB is much more successful than MO theory; however, in general VB theory is not as simple to implement computationally as MO theory, and also suffers the handicap of being less readily available in standard computer packages.

Simple Molecular Orbital Theory for Second Row Homonuclear Diatomics

The valence *ao's* of elements Li to Ne may be taken as *2s, $2p_x$, $2p_y$* and *$2p_z$*, though any independent linear combination of the *2p* orbitals will suffice

(Fig.1.15). Let us begin by considering only these orbitals. As noted above, there are two rules for the degree of mixing of *ao's* to make *mo's*:
 (i) it is inversely related to the energy separation between the orbitals, and
 (ii) it is proportional to the net overlap of the orbitals.
When we later come to mix approximate *mo's* to form better ones, then it is also helpful to know that two *mo's* will have no net overlap if the orbitals being mixed have different symmetries.

In order to determine the diatomic MO energy level diagram, let us firstly assume there is no mixing of s with p orbitals (*i.e.* ignore the $2s/2p_z$ overlap). Then for the homonuclear diatomic M_2, the MO energy level diagram of Fig. 1.17 results. The sketches indicate where the electron density would be if the orbitals were occupied. $2s$ orbitals mix as did the *1s* orbitals for H_2. $2p_x$ only overlaps with the $2p_x$ orbital on the other atom making a bonding and antibonding orbital. Similarly, for the $2p_y$'s and $2p_z$'s, although the nature of the overlap differs for $2p_z$. The labels given to the orbitals are symmetry labels for the $D_{\infty h}$ point group, (see §2.1.2) with the orbitals of the same symmetry numbered consecutively beginning with the *1s* core orbitals (omitted from this diagram). σ means that if the orbital is viewed down the z (bond) axis, then it looks like an s orbital, similarly π orbitals look like p orbitals when viewed down that axis.

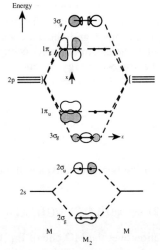

Fig. 1.17 Second row diatomic MO energy level diagram in the absence of *s/p* mixing.

One of the fundamental theorems in quantum mechanics (called the virial theorem [15]) states that any approximate wavefunction will be of higher energy than the true wavefunction. So, we may improve (*i.e.* take it closer to an accurate description of the behaviour of the electrons) the *mo's* by mixing together more *ao's*; at worst the new *mo's* will be the same as the old. The simplest additional mixing is to acknowledge that a $2s$ *ao* on one M will overlap with the $2p_z$ *ao* the other M. Rather than redoing the analysis from the beginning, we may take a shortcut and combine the no *s/p* mixing *mo's* of Fig. 1.17 to make new *mo's*. As noted above, only mixing orbitals of the same symmetry will have any net effect. Thus the two σ_g orbitals mix to give a bonding and an antibonding combination, similarly the two σ_u orbitals mix resulting in the diagrams of Fig. 1.18. The differences between Fig. 1.18 a and b result from different degrees of mixing. $2\sigma_g$ may now be described as bonding/bonding, but $3\sigma_g$ is antibonding/bonding etc.

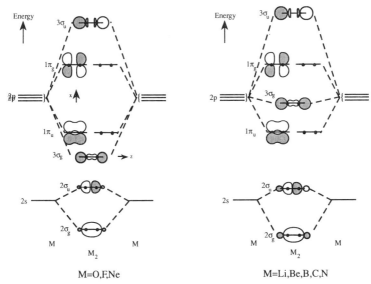

Fig. 1.18 MO diagram for M_2 including s/p mixing. The changes in the orbital shape are exaggerated.

There remains the question of which diagram is appropriate for a given molecule? The usual answer follows from the observation that the mixing of orbitals to make *mo's* is approximately inversely proportional to their energy difference. Atoms on the right hand side of the periodic table, *i.e.* O, F and Ne, have greater nuclear charge than those of the left hand side and so have their electrons attracted more strongly to the nucleus. This exaggerates the energetic difference caused by the fact that on average $2s$ electrons are closer to the nucleus than $2p$ electrons. Thus, although both s and p electrons are lower in energy on the right, the s/p gap is actually bigger (Fig. 1.19) and so less mixing occurs. Thus Fig. 1.18a relates to the left hand side and 1.18b to the right hand side of the periodic table.

The MO diagrams for the second row homonuclear diatomics are given in Fig. 1.19; these have been obtained from calculations that are accurate within the assumption that a single orbital may be used to describe the behaviour of each electron. The occupied valence *mo's* up to C_2 conform with Fig. 1.18a. $3\sigma_g$ and $1\pi_u$ of N_2 are known to be inverted in such calculations with respect to the order observed in photoelectron spectroscopy. In any case they are exceedingly close. The calculations for O_2 and F_2 do not follow Fig. 1.18b, though that for Ne_2 does. At least part of the reason for this lies in the fact that the π_u and π_g orbitals are not simply a combination of $2p$ orbitals as is shown by their asymmetric distribution about the *ao* energies; higher energy orbitals of the appropriate symmetry were included in the calculations.

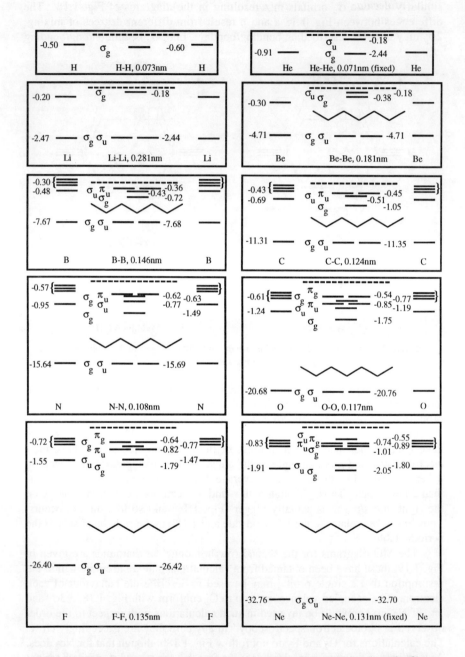

Fig. 1.19 Energies of occupied *mo's* for first and second row homonuclear diatomics from Hartree-Fock calculations using a 6-31G* basis set. Atomic energy levels are also included for comparison. The dotted line is zero. The levels below the zig-zag have an energy scale 1/4 that above, except for Be$_2$ where the factor is 1/2.

Simple Molecular Orbital Theory for Second Row Heteronuclear Diatomics

The step from homonuclear to heteronuclear diatomics is made by remembering the rules governing the degree of mixing. The difference in electronegativities between atoms determines the energy separation of orbitals, thus ultimately determining the direction of electron flow. The labelling of the orbitals also changes due to the reduction in symmetry from $D_{\infty h}$ to $C_{\infty v}$ (see Appendix 2). Some representative diagrams are given in Fig. 1.20.

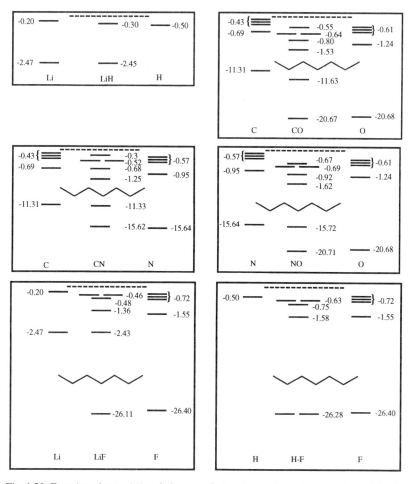

Fig. 1.20 Energies of occupied *mo's* for some first and second row heteronuclear diatomics determined as for Fig. 1.19. The dotted line is zero. The levels below the zig-zag have an energy scale 1/4 that above.

Concluding Comments

MO theory has been used widely in the study of bonded main group systems. However, for non-bonded interactions, such as in van der Waals complexes, and for many electron nuclei, such as transition metals, it has been more problematical. This has not prevented accurate MO calculations being performed but the computer time required is at the forefront of technology.[20] A wide variety of other semi-empirical MO approaches have been developed to endeavour to treat systems with minimum effort for maximum accuracy. A good starting point for exploring these is reference [21].

In this book we are not concerned with performing accurate calculations, but with understanding principles that determine molecular geometry. Simple MO ideas are ideal for understanding what holds a diatomic molecule together. However, this clear picture is lost when we try to make a rigorous extension to polyatomic molecules; this is largely because MO theory allows each electron to delocalise over the whole molecule in a manner not in accord with chemical intuition. For example, the MO picture of NH_3 involves *mo's* that combine all three H atoms together. On the other hand we are taught as chemists to speak of hybridising the $2s$ and $2p$ nitrogen *ao's* to form four orthogonal (*i.e.* non-overlapping) orbitals, three of which will interact separately with an H to form distinct bonds. This raises the important question of whether we really can apply MO theory to selected bits of a molecule in order to understand individual bonds within molecules, or must MO theory only be used for the molecule in its entirety?

Many workers have invested considerable effort in resolving this dilemma by developing methods of localising *mo's*. One of the most successful for this is that of Hansen and Bouman.[22] These processes make it possible (but not easy) to use MO theory in the bond oriented manner demanded by chemical intuition. Recently, Schipper has taken another route and shown how MO theory and the chemist's ideas of distinct chemical bonds can be melded together. The basis of this justification is beyond the scope of this book and the interested reader is referred to [23] for full details. A similar justification has been developed by Burdett in his "fragment formalism".[24] The important point for our puposes is that we are justified in taking both the simplicity of MO theory and the conviction that bonds are real and useful concepts, and then apply MO theory to bits of molecules, such as the various M-L bonds in a metal complex, without having to treat the full complexities of the molecular environment.

1.3.2 Steric-Plus-Electronic Methods

The focus of both MO and VB theory is on what the electrons are doing under the influence of the nuclei. Geometry determination using them proceeds by performing a calculation at a given geometry and then identifying geometry changes that lower the energy. Many other approaches to molecular geometry have a similar approach, but without asking what is making the bond between atoms: the quantum mechanical nature of the electrons is ignored and the molecular formula is taken as given. An approach that completely ignores the electrons except in so far as they contribute to the size of the atoms may be loosely referred to as a steric model for molecular geometry. Often, electronic considerations will be tacked on to the end of such theories in an attempt to make the model semi-quantitative. A number of steric or steric-plus-electronic models for molecular geometry will be considered in the remainder of this chapter.

Let us begin by considering a hypothetical stepwise formation of the molecule ML_n from its component parts. Our notation here is inspired by the notation of metal complex chemistry but is not limited to that field; M denotes the central atom that is usually more electropositive than L but need not be a metal, and L denotes the ligand atom (or group of atoms) that is bonded to M though not necessarily using d orbtials to do so. The electron density of an isolated M atom is spherical. When a ligand L approaches closely enough to M to interact, M's electron density is distorted so as to ensure that there is enough electron density in the M-L bond region to form a bond (usually some of that electron density will be provided by M and some by L), and the remaining electron density relaxes to its most stable arrangement. This distortion and relaxation of the electrons around M will then be repeated with each successive addition of an L.

The energy of the final system can be approximated by the sum of three contributions: M-L interactions; L-L interactions; and the distortion of M's electron density. We shall call these interactions electronic, steric, and stereoelectronic - using this label is the most general sense of geometry dependent electronic effects (see §3.2.2) - respectively. The final geometry will result from an interplay between these three interactions, but one can normally discern a hierarchy of importance such that the M-L interactions are usually the strongest, and the stereoelectronic ones the weakest (though they may still be geometry determining as we shall see). The various (semi-) empirical approaches to molecular geometry differ in the way they treat one or more of these three interactions. Generally, they focus on the first two interactions; in some instances, however, the strength of the electronic interactions is very geometry dependent and stereoelectronic effects cannot be neglected. A good example is the anomeric effect in sugars (see §3.2.2). To a first approximation we shall ignore all stereoelectronic effects. The two questions that need to be answered in order to determine the geometry of ML_n are then: how is the number of L about M determined (this is called the coordination number, C_N), and (ii) how are the L are arranged about M?

Determining Coordination Number

The most stable ML_n system will have as many ligands as there are electrons to make stable bonds, subject to the constraint that the L must fit around M with appropriate bond lengths to ensure good M-L bonds. (In practice second row M have a maximum of four L, third and fourth row M a maximum of six, and larger M seldom have more than eight or nine.) For most molecules, the M-L interactions dominate and the molecule may be viewed as optimising this interaction first, and treating the L-L and stereoelectronic effects as a minor perturbation to the M-L interaction. Bond lengths are very well defined in such systems, and the value of the bond length shows no significant variation if its molecular environment is changed (*e.g.* by changing the other substituents bonded to M).[25-27]

Sometimes the number of ligands is fewer than the sizes of M and L suggest. There are a number of reasons for this: (i) too few valence electrons; (ii) too many valence electrons; (iii) an inability to use valence electrons for bonding; or (iv) unusually strongly repulsive L-L interactions.

(i) There just may not be enough valence electrons to act as glue between M and the maximum number of L. The norm is two-centre two-electron bonds between M and each L (though the whole of borane chemistry is an exception to this rule, §3.3). This deficiency is particularly common where M comes from the left hand side of the periodic table. In some instances L provides the electrons necessary to make a bond so that the maximum number of L may be accommodated; however this is only common with transition metals.

(ii) The *mo's* of a molecule may be classified as bonding (electrons in these orbitals hold atoms together and stabilise the system), non-bonding (electrons in these orbitals have more-or-less the same energy as their constituent atomic orbitals), or anti-bonding (electrons in these orbitals favour the atoms moving apart and make a destabilising contribution to the energy of the molecule). If possible, the system will avoid occupying anti-bonding orbitals. Although core orbitals may be described as bonding and antibonding, they are effectively non-bonding (*cf.* Fig. 1.19-20), so we are most interested in the valence orbitals, *i.e.* those made largely of the valence *ao's*. As a general rule, second row atoms have four valence orbitals, one $2s$ and three $2p$, that make at most four bonding orbitals, and so there is a maximum electron count of eight about a second row atom in a stable molecule. This is known as the eight electron rule and is part of the reason for the C_N of second row elements being four or less. The argument is less specific for third row atoms as the $3d$ orbitals are valence orbitals but of higher energy than the s and p orbitals so may or may not be used. Hence a maximum of somewhere between eight and eighteen electrons might be expected for third row atoms. (*cf.* Figs. 1.14-15 for pictures of orbitals.) If the sterically allowed number of ligands means there will be more than eight or eighteen electrons present then the molecule is unlikely to form (though see §5.1.4 for exceptions where non-bonding electrons increase the count to twenty-two).

(iii) In order to use its electrons in a bond an atom must make them available in the bonding region of space. We may describe how the atom does this by using the concept of a hybrid *ao*, or *ho*, which is typically a mixture of s and p character, but may have d character if there are available d valence *ao's*. Consider carbon: its valence orbitals are $2s$ and $2p$. The latter are oriented to make three bonds at 90° to one another and the $2s$ orbital is not ideally suited for any bonding as it is doubly occupied (so not available to share any of the ligand electrons) and spherically distributed around the carbon (so not localised in space for a bond). For the price of exciting one of the $2s$ electrons into the unoccupied $2p$ orbital, the atom's total electron distribution can be subdivided into four orbitals that have 3/4 p character and 1/4 s character and are oriented tetrahedrally with respect to one another. The radial functions for the *s/p ho's,* are illustrated in Fig. 1.21. Note that an *ho* does not have just one lobe and nor does it have a node precisely at the atom. Carbon is now ready to make its four bonds, and the energy price for the hybridisation is compensated by the four stronger bonds it can now make. Further down Group 14, however, the bond energy is not sufficient to pay the hybridisation price and molecules such as $SnCl_2$ are found. Such pairs of electrons are referred to as "stereochemically inactive" (see Chapter 4).

(iv) In some situations the repulsion between adjacent L atoms can be much stronger than one would expect from the size of L. This is particularly relevant where the M-L bond is polar, since this will lead to an excess of charge on each

L, and hence to an electrostatic repulsion between the L. This is well known in intermolecular interactions and is, for example, one of the main causes of the expansion of water at low temperatures; although less common, it does also occur in intramolecular systems.

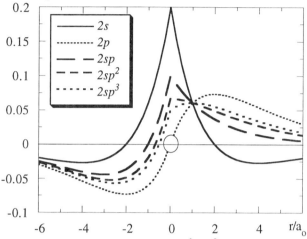

Fig. 1.21 Radial distribution, $R(r)$, of hybrid ao's: $2sp^3$, $2sp^2$, and $2sp$ compared with that of $2s$ and $2p$ (cf. Fig. 1.14 and §1.2 for definition of R).

Arrangement of Ligands about the Central Atom

Once the number of L about M, n, has been established, the remaining question is what positions do the L adopt. All the purely steric approaches to molecular geometry are based on the premise that electronic factors determine n and may then be ignored, while steric factors determine the orientation of the L about M. This is by no means always the case, especially for transition metal systems, but it is still helpful to proceed with this division and bring in "stereoelectronic" factors as a perturbation. The main types of steric geometry models are epitomised by the non-bonded radii approach, the valence shell electron pair repulsion approach, and the atom-atom interaction model. These are outlined below and referred to throughout the remainder of the book. A description of molecular mechanics, which is probably best described as the computational implementation of any non-quantum mechanical geometry determining method, concludes the chapter.

Non-Bonded Radii Approach: The simplest steric approach to molecular geometry was developed by Bartell and extended by Glidewell.[28,29] Bartell noted that for a wide range of main group atoms on the right hand side of the periodic table (*i.e.* non-ionic compounds) the distance between non-bonded atoms showed less variation than that between bonded atoms. He therefore postulated that packing atoms at their non-bonded diameter (approximately the sum of their van der Waals radii, Table 1.1) determined the geometry of molecules. Neither Bartell nor Glidewell explained why their approach worked, but it is remarkably successful. The approach requires that the M-L bond strength be independent of the relative orientations of the L, and that the L are attracted to one another until they come into contact. All details of electronic interactions are hidden in the values of the bond lengths, and these are usually obtained experimentally.

Table 1.1. The non-bonded radii of Bartell[28] and Glidewell[29] (in pm) determined assuming that covalently bound geometries are determined by close packing.

					H = 108
Be = 139	B = 133	C = 125	N = 114	O = 113	F = 108
	Al = 185	Si = 155	P = 146	S = 145	Cl = 144
		Ge = 158		Se = 158	
		Sn = 188	Sb = 188	Te = 187	

Valence Shell Electron Pair Repulsion Theory: Valence Shell Electron Pair Repulsion Theory (VSEPR)[30] takes as different a view from that of the non-bonded radii approach for the adoption of a given geometry as is possible within a non-quantum framework. Gillespie and Nyholm developed the theory soon after Lewis published his concept of the arrangement of electrons in pairs in a molecule. VSEPR begins with the assumption that once the bonds have been formed, the next largest energetic contribution to molecular stability is to place non-bonding electrons in lone electron pairs localised in space. Thus it is stereoelectronic factors that are seen to dominate molecular geometry determination in this model.

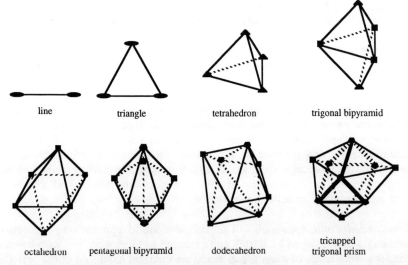

Fig. 1.22 The polyhedra adopted by electron pairs in VSEPR theory. Vertices are represented by shapes indicating the number of edges meeting there.

The geometry is then determined by minimising the steric repulsion between the pairs of electrons, both lone and bonding. The result is that ligands and lone pairs adopt positions that keep them as far from each other as possible; this puts them at the vertices of special geometric shapes (Fig. 1.22). If there is more than one kind of electron pair the hierarchy of repulsion is:

lone pairs > triple bonds > double bonds > single bonds.

Lone pairs are most repulsive as they are contracted towards the nucleus and so occupy a greater solid angle than bonding pairs. Furthermore, electronegative ligands require less space as they draw the electron density away from the nucleus so a smaller solid angle is required to accommodate them. In an odd electron system, a lone electron occupies less room than some pairs.

Many of the molecular examples given in the succeeding chapters illustrate the success of the theory in accounting for observed molecular geometries, though its failures suggest that it is in some way deficient. Perhaps its greatest strengths are that it is easy to apply and its failures are easy to remember, so at the very least it is a valuable mnemonic for remembering molecular geometries. Some examples are illustrated in Fig. 1.23.

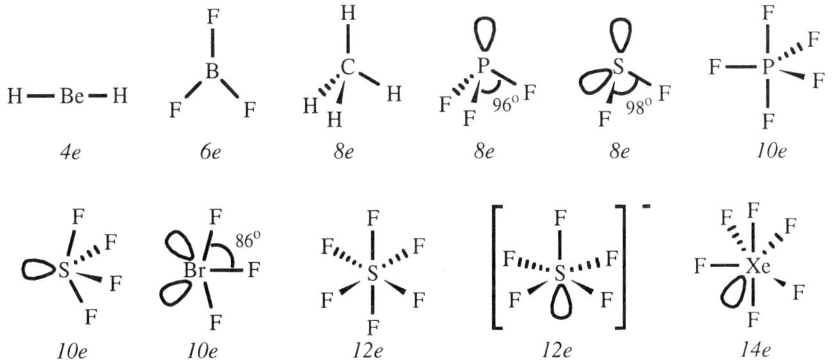

Fig. 1.23 Applications of VSEPR theory. Note *e.g.* equatorial positions of a trigonal bipyramid have more room than the axial ones.

VSEPR fails for very ionic systems such as Li_2O, which is probably more correctly thought of as $Li^+{}_2O^{2-}$, with repulsion of the positively charged lithium atoms rather than of bonding electron pairs dominating. In addition, VSEPR in the form stated above has difficulty with many third and fourth row elements (*cf.* Chapter 4). For example, it cannot account for the square pyramidal geometry of $SbPh_5$, or the bent geometry of SrF_2, and significant problems arise with transition metals. In an excellent book on the VSEPR model[31] Gillespie and Hargittai considered the effects on geometry of the polarisability of the valence cloud of electrons to account for these problems. They thus allow relaxation of the dominating influence of localised lone pair formation on molecular geometry by allowing the lone pair to spread out around the core and so push the bonding pairs together. The intriguing thing about this is that in so doing the difference between the VSEPR approach based on electron pair repulsion and those based on L-L interactions is almost removed. The atom-atom interaction model (AAIM, see below) would consider the driving force for this lone pair spreading as the attraction of the L atoms.

The Atom-Atom Interaction Model (AAIM): The AAIM is based on the assumption that, once the M-L bonds have formed, the L-L interactions determine the orientation of the L's around M in ML_n. Underlying this statement is the assumption, also used in VSEPR theory, that M-L bond strength is independent of orientation. This is not entirely valid; the strength of an M-L bond will change with geometry if there are associated variations in the bonding electron density. This problem is particularly pertinent in "electron deficient" systems such as boron. BH_3 is calculated to be planar,[20] yet the L-L interaction (see below) would favour a pyramidal geometry such as is observed in NH_3. In BH_3 a pyramidal geometry with two-electron bonds would require all the valence shell electron density to be concentrated on one side of the B, whereas in the planar arrangement a high electron density in the bonding region is more compatible with an even distribution around the B atom. This is an example where stereoelectronic considerations would outweigh the steric factors. It is interesting to note that there is no experimental evidence for the existence of BH_3 as a stable species - it always dimerises to form B_2H_6, thus enabling both atom-atom attraction and reasonable distribution of the electrons. Despite such exceptions, it is still useful to begin by assuming the M-L bond strength is independent of orientation and to consider orientational effects as an additional stereoelectronic perturbation where necessary.

In general, the L-L interactions will consist of both repulsive and attractive contributions. The repulsive term is mainly a very strong but short ranged interaction that arises when two atoms or molecules try to occupy the same space; they correlate with our ideas of atomic size and steric hindrance. These interactions are often referred to as excluded volume effects, hard-core repulsion, or overlap forces, although the last should not be confused with the concept of "overlap" as used above in discussing bonding. The attractive term is a softer effect and is active over longer distances than are the overlap forces. The most important contribution is the so-called van der Waals attraction, which is a universal force of attraction between all molecules or atoms and arises from subtle correlations between the motion of the electrons on each atom. In addition there can be electrostatic interactions between the L, and these can be either attractive or repulsive depending on the nature of the L.

A number of attempts to rationalise molecular geometry have been developed by concentrating entirely on the overlap repulsion, the work of Kepert being a notable example (see §5.1.5). These are often quite successful as so many systems have the L closely-packed, and so the excluded volume of each L will be important in determining just how they can fit together. However, in molecules where the bonding is less crowded, the intermediate and long-range L-L interactions come into play. The molecular extreme of this is provided by the van der Waals complexes formed when two or more molecular species associate. Geometries of such species can be modelled very well simply on the basis of the size (*i.e.* the excluded volume) of all the atoms involved, and a good model of the electrostatic interactions between the molecular components.[32,33]

Definition and Determination of Molecular Geometry

In order to become more quantitative it is necessary to find functions that will describe the strength of the different interactions, and in particular, how this varies with the distance between the atoms. Traditionally this is done as a series of r^{-n} terms, with the range of the interaction determining the value of n; as n increases the range over which the interaction is significant becomes smaller. For two of the types of interaction cited above there are rigorous theoretical grounds for choosing this r-dependence: electrostatic interactions can be associated with the usual Coulombic charge-charge interaction and so will vary as r^{-1}; van der Waals forces, also called dispersion forces, vary as r^{-6}. The hard-core repulsion is usually described with an r^{-12} dependence, although this is chosen for pragmatic reasons and other functions are possible; the main requirement is that this term should dominate the interaction energy at small atom-atom distances so as to ensure that the model cannot give rise to cold fusion! The total L-L energy comes from adding these three types of interaction for each distinct atom pair ij at distance r_{ij} from one another

$$E(\text{L-L}) = \Sigma_{ij}\ c_{12}r_{ij}^{-12} - c_6 r_{ij}^{-6} + c_1 r_{ij}^{-1} \quad (1.4)$$

If the L are charged, either intrinsically or due to an ionic or polar bond to M inducing a partial charge, then the charge interaction may become a significant factor in determining the geometry of a system. Usually it is repulsive and tends to push the L apart, thus opposing the effect of the dispersion interaction and taking the atoms towards the same polyhedra as those that minimise electron pair repulsion for VSEPR theory; however, it can be attractive for mixed ligand systems.

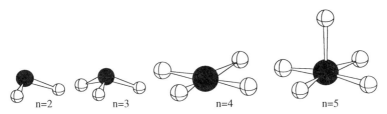

Fig. 1.24 Some dispersion favoured geometries, for non-close packed systems.

The AAIM has been applied successfully to a wide range of molecules, and has been particularly useful in the study of transition metal complexes, as discussed in §5.1.5. An advantage of the model that is not immediately obvious is that it may be applied to the geometries of transition states as profitably as to ground state systems. However, one feature it cannot in anyway account for is the fact that bond angles of significantly less than 90° are observed only rarely (examples include cyclopropane).

If L is uncharged, then the dispersion energy is the dominant L-L attraction. The dispersion energy due to the interaction of two (hard sphere) ligands is approximately [34]

$$E_{disp}(\text{L}_1\text{-L}_2) \approx -3/2\ E_{12}\ \alpha'_1 \alpha'_2\ r_{12}^{-6} \quad (1.5)$$

where α'_i is the polarisability of L_i, and $E_{12} = E_1E_2/(E_1+E_2)$ where E_i is an average excitation energy of L_i (the ionisation energy of L_i is usually a sufficiently good value for E_i). Thus for the total L-L interaction to be maximally stabilising, all L's are brought as close together as possible – which is precisely the non-bonded radii approach. The dispersion interaction for small ligands favours the close packed shapes illustrated in Fig. 1.24. The observed geometry is then a balance between the dispersion (*i.e.* L-L attraction) favoured geometry and the charge (*i.e.* L-L repulsion) favoured geometry. The dipoles and polarisabilities of Table 1.2 enable an estimate of their relative importance to be made.

Table 1.2: Some volume polarisabilities, α' [35-37] and dipoles, μ.[38] q is the partial charge on the atoms. q is determined from the dipole by dividing it by the bond length;[11,19] note that 1D = 480.298e pm.

compound	μ / D	q / e	fragment	α' / Å^3
HF	1.91	0.43	F$^-$	0.96
HCl	1.08	0.18	Cl$^-$	3.60
HBr	0.79	0.12	Br$^-$	5.0
HI	0.38	0.049	I$^-$/I	7.6/4.96
LiH	5.828	0.76	Li$^+$/Li	0.03/22
LiF	6.38	0.84	H$^-$	10.18
LiCl	7.08	0.729	He	0.20
NaCl	8.97	0.79	Na$^+$/Na	0.19/21.5
ClF	0.85	0.108	Ne	0.40
BH	1.27	0.22	Cs$^+$/Cs	2.6/42.0
BF	0.5	0.079	CH$_2$	2.099
AlF	1.53	0.195	O^{2-}	2.74
CO	0.112	0.021	S^{2-}	8.94
CS	1.97	0.27	CO	1.95
SiO	3.09	0.56	CN$^-$	3.47
SiS	1.74	0.17	OH$^-$	1.95

Molecular Mechanics

Underlying all of the methods mentioned so far has been the idea that geometry results from the molecule adopting the lowest energy structure available. When you follow this idea through it leads to the concept of finding minima on a complicated PE surface that describes how the energy of the molecule changes as its constituent atoms move. Methods such as the non-bonded radii approach, VSEPR and AAIM have then proceeded by ignoring certain contributions to the energy and seeking to understand the trends implied by what is left. In the case of AAIM, we then began to see how the model could be made quantitative so that it should be possible to predict precise structures; this involved finding a functional form that would give an adequate description of each (relevant) contribution to the total energy.

In fact, there is no reason why we should not extend this approach and seek a method that will describe the intramolecular potential energy in its entirety. The rigorous way to achieve such a task would be via accurate *ab initio* quantum mechanical calculations, but as we have already mentioned, our ability to do such calculations is limited to relatively small molecules. An alternative process, which forms the basis of the Molecular Mechanics (MM) method, is to find an empirical description of the potential energy surface.[39] In order to provide flexibility to the model and to be able to use MM predictively, it is important that the empirical potential be based on identifiable properties that can be transferred between different system - often referred to as "transferable parameters". The MM potential is also often referred to as a force field.

The MM approach may also be described in terms of electronic, steric, and stereoelectronic factors. The method again begins with the electronic effects since these are usually the strongest interaction, but unlike the earlier approaches MM seeks to quantify the energetics of bonding. Traditionally, these are modelled by identifying ideal bond lengths, bond angles and torsion angles (*i.e.* bond twists), since these show little variation over a wide range of different molecules, and then determining the energetic penalty involved in distorting away from these ideal values. Thus the model consists of a set of ideal geometric properties, and the force constants that quantify the restoring force that arises when the atoms move. Steric forces are then included in the same way as for the AAIM,[†] so that a typical MM potential energy function would take the form

$$E = E(\text{bond length}) + E(\text{bond angle}) + E(\text{torsion}) + E(\text{hard-core repulsion})$$
$$+ E(\text{van der Waals}) + E(\text{electrostatic}) \qquad (1.6)$$

It is also worth mentioning that the MM method is not synonymous with the form of the potential energy given in Eq. 1.6. In fact the method is far more general, and should be associated with any attempt to minimise energy using a potential energy function that describes both electronic and steric interactions and is based transferable parameters; the actual description chosen for the potential will obviously affect the accuracy of the method, but is not intrinsic to the method itself.

The one thing that has been missing from this discussion of MM is stereoelectronic effects, and it is not easy to identify these within MM. In a crude sense they will be hidden in the choice of parameters, particularly for the bond angle terms. Thus in choosing ideal bond angles of 90° or 109° about a central metal atom in a metal complex, one is making a statement about the way the bonding has distorted the electron distribution about the metal. Further, since the potential parameters in MM tend to be chosen empirically (*i.e.* to agree with experiment) and stereoelectronic effects will be present in the experimental data, the MM potential will incorporate such effects in some average sense. However,

---

[†] Note that in the MM literature, non-bonding effects (excluding electrostatic terms) are often known collectively as van der Waals energies. This is not strictly correct, since this term should really apply only to the r^{-6} attractive dispersion interaction.

the intrinsic nature of stereoelectronic effects is that they are environment-dependent, and so any accurate description must run counter to the idea of transferable parameters that underlies the success of MM.

MM is a very general method, and has the power to give accurate predictions of molecular geometry even in quite complicated systems. Typical MM force fields have been designed to reproduce a range of experimental properties for small and moderate sized molecules, including detailed structures, energy differences between different isomers, vibrational frequencies and torsional energy barriers.[40] The method has had considerable success with organic compounds,[41-43] and is now developing into a reliable method for inorganic and organometallic systems as well.[44-46] Unfortunately, the generality of the method is also its main drawback, and one cannot hope to implement MM for anything but moderately simple molecules without access to reasonably powerful computer facilities. Further, the increased complexity of the potential energy function means that one has to work a lot harder to extract the physical principles behind the molecular geometry. For these reasons, it is unlikely that MM will ever remove the need for the simpler models such as non-bonded radii, VSEPR and AAIM.

References

(1) Bailar, J. C.; Emeleus, H. J.; Nyholm, R.; Trotman-Dickenson, A. F. *Comprehensive Inorganic Chemistry*; Pergamon Press: Oxford, 1973.
(2) Cotton, A. F.; Wilkinson, G. *Advanced Inorganic Chemistry;* 5th ed.; Wiley-Interscience: New York, 1988.
(3) Greenwood, N. N.; Earnshaw, A. *Chemistry of the Elements;* 1st ed.; Pergamon Press: Oxford, 1984.
(4) Huheey, J. E. *Inorganic Chemistry: Principles of Structure and Reactivity;* 3rd ed.; Harper International: New York, 1983.
(5) Porterfield, W. W. *Inorganic Chemistry: A Unified Approach*; Addison-Wesley Pub. Co: U.S.A., 1984.
(6) Wells, A. F. *Structural Inorganic Chemistry;* 4th ed.; *Clarendon Press*: Oxford, 1975.
(7) Trueblood, K. N.; Dunitz, J. D. *Acta Cryst.* **1983**, *B39*, 120.
(8) Bond, E.; Muetterties, E. L. *Chem. Rev.* **1978**, *78*, 639.
(9) Longuet-Higgins, H. C. *Molec. Phys.* **1963**, *6*, 445.
(10) Eaton, S. S.; Hutchinson, J. R.; Holm, R. H.; Muetterties, E. L. *J. Amer. Chem. Soc.* **1972**, *94*, 6411.
(11) Aylward, G. H.; Findlay, T. J. V. *SI Chemical Data*; John Wiley and Sons, Australasia Pty. Ltd.: Sydney, 1972.
(12) Rodger, A.; Schipper, P. E. *Chem. Phys.* **1986**, *107*, 329.
(13) Kepert, D. L. *Prog. in Inorg. Chem.* **1979**, *23*, 1 and refs. therein.
(14) Berry, R. S., Rice, S. A.; Ross, J. *Physical Chemistry*, John Wiley and Sons: New York, 1980.
(15) Eyring, H.; Walter, J.; Kimball, G. E. *Quantum Chemistry*; John Wiley and Sons: New York, 1944.
(16) Atkins, P. W. *Molecular Quantum Mechanics*; Oxford University press: Oxford, 1983.
(17) Gimarc, B. M. *Molecular Structure and Bonding: The Qualitative Molecular Orbital Approach*; Academic Press: USA, 1979.
(18) McWeeney, R.; Sutcliffe, B. T. *Methods of Molecular Quantum Mechanics*; Academic Press: London and New York, 1976.
(19) Levin, G.; Goddard, W.A. *J. Amer. Chem. Soc.* **1975**, *97*, 1649.

(20) Hehre, W. J.; Radom, L.; Schleyer, P. v. R.; Pople, J. A. *Ab Initio Molecular Orbital Theory*; John Wiley and Sons: New York, 1986.
(21) Clark, T. *A Handbook of Computational Chemistry*; Wiley-Interscience: New York, 1985.
(22) Lightner, D. A.; Bouman, T. D.; Wijekoon, W. M. D.; Hansen, A. E. *J. Amer. Chem. Soc.* **1986**, *108*, 4484.
(23) Schipper, P. E. *J. Phys. Chem.* **1986**, *90*, 2351; Schipper, P. E. *Recueil des Travaux Chimiques des Pays-Bas* **1990**, *109*, 1.
(24) Burdett, J. K. *Moleculer Shapes: Theoretical Models of Inorganic Stereochemistry*; John Wiley and Sons: New York, 1980.
(25) Orpen, A. G.; Brammer, L.; Allen, F. H.; Kennard, O.; Watson, D. G.; Taylor, R. *J.C.S. Dalton Trans* **1989**, S1.
(26) Allen, F. H.; Kennard, O.; Watson, D. G.; Brammer, L.; Orpen, A. G.; Taylor, R. *J.C.S. Perkin Trans II* **1987**, S1.
(27) Bowen, H. J. M.; Donohue, J.; Jenkin, D. G.; Kennard, O.; Wheatley, P. J.; Wiffen, D. H. *Tables of Interatomic Distances and Configuration in Molecules and Ions*; The chemical Society: London, **1958**; Vol. Special Publication no. 11.
(28) Bartell, L. S. *J. Chem. Phys.* **1960**, *32*, 827.
(29) Glidewell, C. *Inorganica Chimica Acta* **1975**, *12*, 219; *Inor. Nucl. Chem.* **1976**, *38*, 669.
(30) Gillespie, R. J.; Nyholm, R. S. **1975**, *11*, 339.
(31) Gillespie, R. J.; Hargittai, I. *The VSEPR Model of Molecular Geometry*; Allyn and Bacon: Boston, 1991.
(32) Buckingham, A. D.; Fowler, P. W. *Can. J. Chem* **1985**, *63*, 2018.
(33) Buckingham, A. D.; Fowler, P. W.; Hutson, J. M. *Chem. Rev.* **1988**, *88*, 963.
(34) Gray, C. G.; Gubbins, K. E. *Theory of Molecular Fluids. 1: Fundamentals*; Clarendon Press: Oxford, 1984; Vol. 1.
(35) Davies, M. *Some Eletricial and Optical Aspects of Molecular Behaviour*; Pergammon Press: Oxford, 1965.
(36) Davies, D. W. *The Theory of the Electric and Magnetic Properties of Molecules*; John Wiley and Sons: 1967.
(37) Bederson, B.; Robinson, E. J. In *Advances in Chemical Physics*; Interscience Pub.: 1966; Vol. 10; pp 1.
(38) Green, S. *Advances in Chemical Physics*, Interscience Pub.: 1974; Vol. 25; pp 179.
(39) Boyd, D. B.; Lipkowitz, K. B. *J. Chem. Ed.* **1982**, *59*, 269.
(40) Momany, F. A.; Rone, R. *J. Comput. Chem.,* **1992**, *13*, 888.
(41) Nilsson, L.; Karplus, M. *J. Comput. Chem.,* **1986**, *7*, 591.
(42) Weiner, S. J.; Kollman, P. A.; Nguyen, D. T.; Case, D. A. *J. Comput. Chem.,* **1986**, *7*, 230.
(43) Jurema, L. W.; Shields, G. C. *J. Comput. Chem.,* **1993**, *14*, 89.
(44) Doman, T. N.; Landis, C. R.; Bosnich, B. *J. Amer. Chem. Soc.,* **1992**, *114*, 7264.
(45) Brunier, T. M.; Drew, M. G. B. D.; Mitchell, P. C. H. *J. Chem.. Soc. Faraday Trans.,* **1992**, *88*, 3225; *Mol. Sci.,* **1992**, *9*, 143.
(46) Hope, A. T. J.; Leng, C. A., Cutlow, C. R. A. *Proc. Roy. Soc. Lond. A*, **1989**, *424*, 57.

CHAPTER 2

A Unified View of Stereochemistry and Stereochemical Changes

Contents

Introduction			35
2.1	Point symmetry		36
	2.1.1	Point symmetry operations	36
	2.1.2	Point symmetry groups - formalism	37
	2.1.3	Chiral and achiral point groups	39
	2.1.4	Examples of point symmetries	41
2.2	Determination of symmetry adapted functions		41
	2.2.1	Molecular orbitals and molecular orbital energy level diagrams from symmetry	41
		Sao's of the π-system of CH_2CHCH_2 (C_{2v})	44
		How many orbitals of each symmetry type: reducible representations	44
		MO energy level diagrams from symmetry	45
		Mo's for H_3 (D_{3h})	46
		Octahedral and tetrahedral point groups	48
		Final comment	54
	2.2.2	Vibrations	54
	2.2.3	Symmetries of wavefunctions	56
2.3	ML_n Geometries and their interconversion		57
	2.3.1	ML_n Symmetry, geometry, and stability	57
	2.3.2	Stereochemical changes	60
	2.3.3	The classical symmetry selection rule procedure	63
		Formalism	64
		Applications	65

Introduction

It is the variety of structures observed in chemical systems that leads to the greatest confusion when one studies molecular geometry. One solution (which the organic chemist is likely to adopt) is to limit consideration essentially to the first and second rows of the periodic table where the geometries are fairly well predicted by a molecular model kit and a little knowledge. However, if that is not the option you wish to take, then some unified view of at least most of the geometries one might encounter is desirable. In §1.1 we saw how the concept of high symmetry templates could be used for studying and systematising molecular geometry. The focus of this chapter is molecular symmetry and we shall develop the template idea further to see how different templates are related to one another and how these relationships can also help us to understand reactions.

Discussions of symmetry are usually quite formalistic. We have limited the need for formalism by making some sweeping statements and putting some of the justifications into Appendices 1 and 2. The interested reader is referred to these at the appropriate place in this chapter. A more detailed discussion may be found in standard textbooks such as references [1-6]. The biggest problem most people have in understanding symmetry is visualising the shapes. For this chapter, a set of cocktail sticks and small pieces of plastic or rubber tubing with holes cut in them or pieces of plasticine is probably the best molecular model kit, as more expensive kits restrict the shapes you may build to what the designer felt was likely to be useful.

The chapter divides into three parts. In the first (§2.1) the symmetry operations of molecules are identified and collected together in point symmetry groups. The second part (§2.2) illustrates the use of projection operators to determine *mo's* (molecular orbitals), vibrations and wavefunctions. The final part (§2.3) examines the relationships between different molecular geometries, both statically and kinetically, concluding with an outline of the "Classical Symmetry Selection Rule Procedure" for determining symmetry allowed reaction pathways. The contents of this chapter are all included because later chapters rely on the concepts, results, and notation. The level of understanding needed of this material depends upon what the reader requires from the later chapters. Some readers may therefore prefer to read this chapter on a "need-to-know" basis in the context of the material of later chapters.

2.1 Point symmetry

2.1.1 Point Symmetry Operations

A symmetry operation may be defined to be some action that when performed on the molecule makes no observable (or measurable) difference. In other words, unless you can label identical atoms, you cannot tell that the operation has happened. The symmetry operations relevant to our study of molecular geometry are the point symmetry operations which leave at least one point of the system unmoved; these operations are defined in Table 2.1. The importance of symmetry operations follows from their definition. If a symmetry operation leaves all observables unchanged, then, if we know the symmetry of a molecule, we know something about its observable properties. In particular we know something about its geometry, energy, and electron density.

By convention, chemists define the direction of rotation to be anticlockwise when the observer is looking down the rotation axis from the positive direction (Fig. 2.1). We use the notation C_n^k to denote C_n repeated k times. So, if C_n is a symmetry operation of a molecule, then C_n^k, must also be one of its symmetry operations. In most molecules there is no more than one C_n operation with $n \geq 3$.

The axis about which this rotation operates is referred to as the major rotation axis, and is denoted z.

Table 2.1 Point symmetry operations.

R_ξ	Name	Description
E	identity	molecule is *unchanged*
C_n	n-fold proper rotation	rotation through an angle $2\pi/n$ about an axis
σ	reflection	reflection in a plane
S_n	improper rotation	C_n followed by σ in plane perpendicular to rotation axis
$i \equiv S_2$	inversion	takes a point at (x,y,z) in Cartesian space to $(-x,-y,-z)$

Reflection planes are labelled with subscripts that describe their position with respect to rotation axes. We refer to reflections in a plane perpendicular to the main rotation axis as horizontal reflections, σ_h; reflections in planes containing the major rotation axis as vertical reflections, σ_v; and vertical reflection planes that also bisect two C_2 axes as dihedral reflections, σ_d. Alternate reflection planes are also referred to as σ_d when there are an even number of σ_v planes in a group.

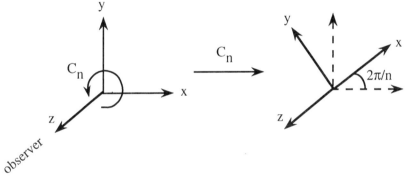

Fig. 2.1 Schematic illustration of the effect of a C_n point symmetry operation. The $\{x,y,z\}$ axes of the molecule (solid lines) are rotated and the external axis system (dotted lines) remains stationary.

2.1.2 Point Symmetry Groups - Formalism

If we start off with a molecule M^0 and operate on it with a symmetry operation R_1 then we produce a molecule $M^1 = R_1\{M^0\}$ that cannot be distinguished experimentally from M^0. So any symmetry operation, R_2, of M^0 is also a symmetry operation of M^1 and we may write

$$M^2 = R_2\{R_1\{M^0\}\} = (R_2 R_1)\{M^0\} = R_2 R_1\{M^0\}$$

where sequential performance of symmetry operations is referred to as multiplication and, by convention, we choose to write the multiplication as proceeding from right to left (*i.e.* perform the right-hand operation first). With this definition of multiplication and the types of operations in Table 2.1, the symmetry operations of a molecule form a mathematical group, as defined below, and are referred to as a point group.

A set of operations $\{R_i, R_j, R_k, ...\}$ form a group if and only if:
(i) There is an operation, denoted E and called the identity, that is a member of $\{R_i, ...\}$ and satisfies: $R_i E = E R_i = R_i$ for all R_i.
(ii) The product of any two operations, $R_j R_k$, is also a member of $\{R_i, ...\}$.
(iii) For any R_i there is another operation R_j that satisfies: $R_i R_j = R_j R_i = E$ (*i.e.* there is an inverse of every operation that undoes the original action).
(iv) Multiplication is associative, *i.e.* $R_i(R_j R_k) = (R_i R_j)R_k$.

CHIRAL

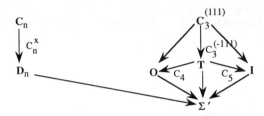

ACHIRAL

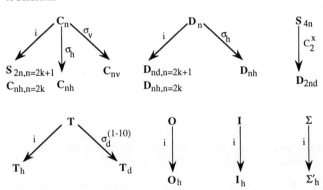

Fig. 2.2 Point group generating tree. Some augmenting operations are indicated beside the arrows. Superscripts indicate rotation axes and normals to reflection planes in Cartesian coordinates.

The reason we bother to note that the point symmetry operations of a molecule form a group is that the mathematical formalism of group theory may then be used to help us apply symmetry systematically to physical problems. If we know all the possible point groups to which molecules may belong, then we have some information about the shapes molecules may adopt. We shall use Schoenflies notation for point groups throughout this book as this is most widely used by chemists. It is discussed in some detail in *e.g.* references [1-6]. We shall denote the point groups with bold faced type, *e.g.* **C**$_n$, and the operations with Roman script, *e.g.* C$_n$. C$_\infty$ is an infinitesimal rotation, *i.e.* what a rotation of $2\pi/n$ becomes as $n \to \infty$. Where necessary a superscript is used to indicate the axis about which a rotation operates. The direction perpendicular to a reflection plane is also indicated by a superscript if the subscripts h, v, and d are not sufficient to identify it.

One consequence of the second property of groups is that one may generate all the operations of a group from a small subset of operations, usually called a *generating* set. No point group has a unique generating set. The labels used to identify most point groups summarise the contents of one of its generating sets. Thus, *e.g.* **C**$_n$ is generated by C$_n$; **C**$_{nh}$ is generated by C$_n$ and σ_h; **D**$_{nd}$ is generated by C$_n$, C$_2$ perpendicular to C$_n$, and σ_d that bisects two of the C$_2$ axes but *does not* contain any of them.

Using generating sets we can develop a hierarchy of the different point groups, as shown in Fig. 2.2, in which new operations are added to the generating set of one group to go to the next higher symmetry group. Such an operation is referred to an an augmenting operation. The effect of multiplying different operations together may be summarised by the eight simple rules given in Appendix 1. The least obvious one, Rule 8, enables us to limit our search for finite point groups with more than one major rotation axis to shapes of the symmetry of the five platonic solids.[†] Appendix 2 describes how to generate all of the point groups.

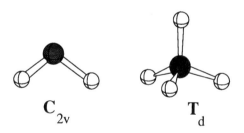

Fig. 2.3 Water and methane.

2.1.3 Chiral and Achiral Point Groups

Point symmetry operations separate into two classes: proper operations or rotations and improper operations or rotations.

A proper operation may be performed without destroying the integrity of the molecule – it simply turns the molecule around. E and C$_n$ are the proper

---

[†] The icosahedron and dodecahedron have the same symmetry as do the cube and the octahedron.

operations. The product of any two proper operations must be another proper operation.

Table 2.2 Minimum set of points required to produce a collection of points (atoms) having a given point symmetry. A set of equivalent points lying on no rotation axis or reflection plane is denoted {a}; a set of points lying on one symmetry element is denoted {b}, where b labels the rotation axis or reflection plane containing the points. 2{a} means two complete sets of type {a} are required to produce the necessary symmetry. A set of points of type {a} is determined by selecting a point in Cartesian space (x, y, z) that does not lie on any symmetry elements (axes or planes) of the group. One then performs the symmetry operations on that point drawing a new point at any place to which a symmetry operation takes the original one.

Point Group	Number of sets of equivalent operations required to generate the point symmetry group
C_n	2{a}
D_n	{a}
T, O, I	{a}
C_{nh}	2{a}, 2{σ_h}, or {a}+{σ_h}
C_{nv}	2{a}, 2{σ}, {σ}+{σ'}, or {a}+{σ} [†]
D_{nh}	{a}, {σ_h}, {σ}, or {σ_h, C_2, σ} [†]
D_{nd}	{a}, or {σ_d}
S_{2n}	2{a}
T_h	{a}, or {σ_h} [††]
T_d	{a}, {σ_d}, or {C_3, 3 of σ_v} [†††]
O_h	{a}, {σ_h}, {σ_d}, {C_2', σ_h}, {C_3, 3 of σ_d} or {C_4, 2 of σ_d, 2 of σ_h}
I_h	{a}, {C_2}, {σ}, {C_3, 3 of σ}, {C_5, 5 of σ}

[†] By convention with the C_{2nv} and D_{2nh} point groups, alternate C_2' axes and vertical planes are denoted C_2' and C_2'', and σ_v and σ_d respectively. We take C_2' and σ_v to contain an atom if possible. Thus for σ read either σ_v or σ_d and for C_2 read C_2' or C_2''.

[††] The reflection planes of T_h are often denoted σ_d; however, we use σ_h as they are the horizontal reflection planes perpendicular to the x, y, z C_2 axes. (N.B. For T point groups, z is a two-fold rotation axis, not three-fold.)

[†††] The reflection planes of T_d are also often denoted σ_d; however, we use σ_v as they contain the C_2 axes.

An improper operation, i, σ or S_n, may not be physically performed on a molecule. Its effect is to invert the handedness of the system. This is readily seen for the inversion, i, which takes $x\rightarrow-x$, $y\rightarrow-y$, and $z\rightarrow-z$. Analogously to negative and positive numbers, the product of two improper (negative) operations must be proper, since two inversions of the handedness of a system have no net effect. Conversely, the product of an improper and a proper operation is always improper. The division of symmetry operations into proper and improper operations underlies the definition of a chiral molecule (*i.e.* one that cannot be superposed on its mirror image, §1.1.1) as a molecule having only proper symmetry operations. The chiral point groups include D_∞ (a chiral "cylinder") together with all its subgroups: C_∞, D_n, and C_n, and **I**, **O**, and **T**. Achiral point groups are most simply generated by augmenting a chiral group with an improper operation σ, i, or S_n that generates no new proper operations (Appendix 2).

2.1.4 Examples of Point Symmetries

It is often easier to think in terms of molecules that belong to specific point groups, such as C_{2v} (water, Fig. 2.3) and T_d (methane, Fig. 2.3). Molecules of many different symmetries are illustrated in the subsequent sections of this book. Some point symmetries, such as **T**, are very unusual in molecules, for reasons that become apparent when the arrangement of points (*i.e.* atoms) that would be required for a molecule to have that symmetry is deduced. Jahn and Teller[7] determined the minimum sets of points required to produce a particular symmetry. Table 2.2 gives their results in terms of sets of equivalent points of different types. Some D_{2d} examples are illustrated in Fig. 2.4. Others can be constructed with a little patience.

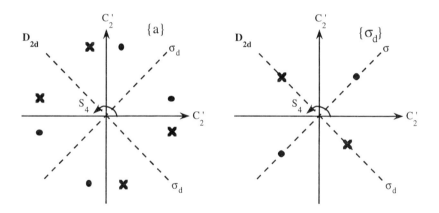

Fig. 2.4 Two sets of points, {a} and {σ_d} that generate D_{2d}. Crosses denote points below the plane of the paper and solid shapes points above.

2.2 Determination of Symmetry Adapted Functions

2.2.1 Molecular orbitals and molecular orbital energy level diagrams from symmetry

In §1.3.1 we used symmetry more-or-less explicitly to help us derive the *mo's* for first and second row diatomics from the valence *ao's* (atomic orbitals) of the constituent atoms. When molecular symmetry is high it is often possible to determine the *mo's* on the basis of symmetry alone. In other cases, the *mo's* must still reflect the molecular symmetry and so it is helpful to form the *ao's* into symmetry adapted orbitals (*sao's*) and then combine the *sao's* to give *mo's*. In either case, we can determine a qualitative MO energy level diagram. The following discussion provides a brief rationale for the use of projection operators to determine *sao's*. The examples that follow illustrate the details of their application. Throughout this discussion we shall use η to represent *ao's*, and ϕ for the *sao's*.

By definition, symmetry operations must leave observables, such as electron density and energy, unchanged, *i.e.* they must be totally symmetric to the operations of the point group. Thus wavefunctions, ψ, and *mo's*, ϕ, which by the Born interpretation are squared to give the observable electron density, are square roots of a totally symmetric quantity.[†] Character tables summarise the ways such square roots can occur, and therefore tell us how to form *sao's*. We do not need to understand how to determine character tables in order to use them, just as one may use a building without having personally inspected the foundations.

We can see how to use character tables by considering how to produce a totally symmetric *sao* from one of the *ao's*, η. If we operate on η with a sum of all the symmetry operations of the point group of the molecule, the result will be a totally symmetric *sao*, since all the non symmetric parts of the wavefunction will have been cancelled out. This sum of symmetry operations is referred to as the totally symmetric projection operator, $P°$.

$$P° = \frac{1}{h}\{E + R_1 + R_2 + ...\} = \frac{1}{h}\Sigma_i\ R_i$$

where the R_i are the point symmetry operations and h is the order of the group (*i.e.* the number of distinct operations in the group). So $P°(\eta)$ is an *sao* that is totally symmetric to all operations of the group.

Now, $P°$ may be written as a scalar product of the vectors $(1,1,1,...)$ and $(E, R_1, R_2, ...)/h$ so

$$P° = \frac{1}{h}\{E + R_1 + R_2 + ...\} = \frac{1}{h}(1,1,...) \cdot (E, R_1, R_2, ...)$$

The totally symmetric *sao* is not the only *sao* that is consistent with a given point symmetry group. Other symmetries are possible, corresponding to different "square roots" of totally symmetric functions. The different ways this can be done are summarised as the rows of the character table for the point group. In the simplest case this just amounts to replacing half of the +1's in the totally symmetric row by -1's in such a way that the product of two operations that have assigned -1 is an operation that has been assigned +1 (*cf.* §2.1.3). These rows of the character table result from the fact that both $(+1)^2 = 1$ and $(-1)^2 = 1$. In more abstract forms of mathematics other square roots of unity can be found, and these also have their counterparts in the character tables. In general, each row of the character table is referred to as a *representation*, and the individual numbers within each row are called *characters*. We are now in a position to define projections operators, P^Γ, that are analogous to $P°$, but are not totally symmetric:

[†] Strictly, $\psi^*\psi$ is the observable, so if ψ is a complex function we have to work a little harder; this will only happen if two or more ψ have the same energy (*i.e.* are degenerate).

A Unified View of Stereochemistry and Stereochemical Changes 43

$$P^\Gamma = \frac{1}{h}\{\chi_E E + \chi_1 R_1 + \chi_2 R_2 + ...\}$$

where χ_i is the character for operation R_i in the Γth row of the character table (see e.g., Table 2.3). P^Γ may also be written in vector notation:

$$P^\Gamma = \frac{1}{h}(\chi_E, \chi_1, \chi_2, ...) \cdot (E, R_1, R_2, ...)$$

The character corresponding to the identity, χ_E, is particularly useful as it specifies the number of different ways this symmetry can be achieved. It is called the *degeneracy* and ultimately specifies the number of *sao's* or *mo's* with the same energy that we need to find. The vectors $\{\chi_E, \chi_1, \chi_2, ...\}$ are orthogonal to (1,1,1,...) and to each other.

If the point group has an operation C_n, $n \geq 3$, but is not C_{nh} or C_n, then some symmetry operations have the same character in *all* representations. In tabulations, therefore, such operations are grouped together into classes and listed in a single column of the character table. A corresponding number of rows of the character tables describe the transformation properties of degenerate functions. In forming the projection operator, however, it is important that the classes be expanded and each operation treated individually.

Table 2.3 The rows of the C_{2v} character table and the effect of C_{2v} symmetry operations on the p_x carbon orbitals of CH_2CHCH_2.

	E	C_2	σ_{xz}	σ_{yz}
A_1	1	1	1	1
A_2	1	1	-1	-1
B_1	1	-1	1	-1
B_2	1	-1	-1	1
p_x^1	p_x^1	$-p_x^3$	p_x^3	$-p_x^1$
p_x^2	p_x^2	$-p_x^2$	p_x^2	$-p_x^2$
p_x^3	p_x^3	$-p_x^1$	p_x^1	$-p_x^3$

Now, to generate *sao's* corresponding to non-totally symmetric representations we need to operate with the relevant projection operator P^Γ on an *ao*, η; this will give the *sao* $P^\Gamma(\eta)$ that has the symmetry properties described by the Γth row of the character table.

$$P^\Gamma = \frac{1}{h}(\chi_E, \chi_1, \chi_2, ...) \cdot (E, R_1, R_2, ...)\eta$$

$$= \frac{1}{h}(\chi_E, \chi_1, \chi_2, ...) \cdot (E(\eta), R_1(\eta), R_2(\eta), ...)$$

If there are two or more non-degenerate *sao's* of the same symmetry produced from the same set of *ao's* we wish to use, then the *mo's* will be some combination of these *sao's* and will have the same symmetry.

The above discussion will become clearer in the context of specific examples and so is illustrated below for C_{2v}, D_{3h}, T_d, and O_h systems.

Sao's of the π-system of CH_2CHCH_2 (C_{2v})

We wish to determine the π *mo's* for CH_2CHCH_2 using a basis set of the $2p_x$ orbitals on the carbons. This system, which has C_{2v} symmetry, is illustrated in Fig. 2.5. The effect of the symmetry operations of C_{2v} on p_x^1, p_x^2, and p_x^3 is listed in Table 2.3, together with the rows of the C_{2v} character table. The first four rows of this table are the $(\chi_E, \chi_1, \chi_2, ...)$ vectors, while the last four correspond to possible vectors $(E, R_1, R_2, ...)\eta$. Any a_1 symmetry *sao's* are found from the scalar products of each p_x orbital with the A_1 row of Table 2.3.[†] Similarly for a_2, b_1, and b_2. For example,

$$P^{A_2}(p_x^1) = (p_x^1 - p_x^3 - p_x^3 + p_x^1)/4$$

After all combinations of Γ and η have been tested, three distinct *sao's* result:

$\phi(a_2) = (p_x^1 - p_x^3)/\sqrt{2}$
$\phi(b_1) = (p_x^1 + p_x^3)/\sqrt{2}$
$\phi(b_2) = p_x^2$

The factors of $1/\sqrt{2}$ have been included instead of the $1/2$ that comes from the projection operator as then ϕ^2 has volume equal to 1 (if the $2p$ orbitals do also). Such wavefunctions are referred to as being normalised. The *sao's* for CH_2CHCH_2 are illustrated in Fig. 2.6. In this case all three functions are both *sao's* and *mo's*, since there is only one *sao* of each symmetry kind (remembering that we are using only the $2p_x$ orbitals to generate *sao's*). It should be noted that the simple rule: *n* functions in, *n* out always works, hence three *ao's* gave three *sao's*. If a larger set of *ao's* were used in the calculation more *sao's* would result. The choice of functions to include in a calculation is referred to as the *basis set*.

How Many Orbitals of Each Symmetry Type: Reducible Representations

If the molecule has many more symmetry operations than C_{2v} then it will also have more rows in the character table. The *sao's* from a chosen basis set normally only come from a few rows of the character table and so it usually saves time if we know which rows are relevant for a given problem. We proceed by determining a vector which describes how many of the basis functions we have chosen are unchanged (giving a count of +1) or inverted (giving a count of -1) by each operation; where the functions have been interchanged, *e.g.* p_x^1 and p_x^3 are exchanged under σ_{xy} in the previous example, then we count zero for both functions. This vector is obtained by tabulating the vectors $(E, R_1, R_2, ...)\eta$ for each η in the basis set, as *per* Table 2.3, and then summing each column; this

---

[†] By convention we use small letters for symmetry labels of orbitals and capitals for the complete wavefunction.

procedure results in a vector known as a *reducible representation*. In the above example the vector is (3,-1,1,-3). We then take the scalar product of this vector with each row of the character table in turn, and divide by the order of the group. The resulting number tells us how many functions of that symmetry we will find. For example, there are (3,-1,1,-3).(1,1,1,1)/4 = 0 a_1 functions in the present example, but (3,-1,1,-3).(1,1,-1,-1)/4 = 1 *sao* of a_2 symmetry. In general, if the answer is zero then there is no point in trying to determine a function of that symmetry from the chosen basis set, while if the answer is 1 then the *sao* is also an *mo*. Under C_{2v} symmetry the vector (3,-1,1,-3) reduces to ($a_2+b_1+b_2$), which are the symmetries of the orbitals determined above.

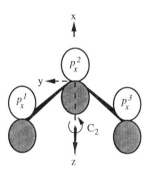

Fig. 2.5 CH_2CHCH_2 illustrating valence π-orbital basis set and coordinate system.

MO Energy Level Diagrams from Symmetry

If two or more *sao's* of the same symmetry are found, then they must be mixed together to produce the *mo's*. The precise degree of mixing is seldom easy to determine without a computer, but it can be estimated qualitatively as we did in Chapter 1 when we admitted the existence of *s/p* mixing; thus it is possible to draw qualitative MO energy level diagrams. If there are two non-degenerate orbitals of the same symmetry, then the net result will be a "bonding" orbital (or in-phase combination) which is more stable than either of the two starting ones, and an "antibonding" orbital (or out-of-phase combination) which is less stable.[†] Three orbitals of the same symmetry usually give a bonding, an anti-bonding and an approximately "non-bonding" orbital.

Determining qualitative *mo* diagrams is often straightforward given the guiding principles that (i) overlap of orbitals is stabilising and (ii) nodes (*i.e.* points at which there is no electron density) between atoms are destabilising. Thus the b_1 orbital of the valence π system basis set of CH_2CHCH_2 illustrated in Fig. 2.6 has the lowest energy, followed by the b_2 orbital. In this example, the ground (lowest) state arrangement of the π electrons has two electrons in the b_1 orbital and one in the b_2 orbital.

[†] The one exception is if the two *sao's* also happen to be the *mo's*, in which case there will be no change in their energies.

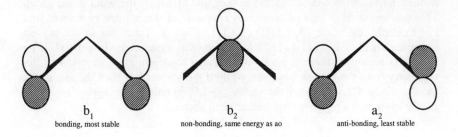

b₁	b₂	a₂
bonding, most stable	non-bonding, same energy as ao	anti-bonding, least stable

Fig. 2.6 CH_2CHCH_2 valence π mo's

Mo's for H_3 (D_{3h})

Groups with n-fold rotations, $n \geq 3$, are referred to as degenerate point groups. They have at least one degenerate representation with $\chi_E \geq 2$. If the above projection procedure indicates that a function belonging to a degenerate representation is present, then there will be precisely χ_E different functions with the same energy that need to be found. The easiest way to do this is usually to repeat the projection on χ_E different but degenerate basis functions. The simplest example of a "molecule" with degenerate *mo's* is the equilateral triangle H_3. In this case the valence orbitals we shall use as basis functions are the *1s* orbitals on each H, which we denote $s^1 = \mathbf{1}$, $s^2 = \mathbf{2}$, and $s^3 = \mathbf{3}$ respectively. The transformation vectors for these three orbitals are summarised in Table 2.4 together with the $\mathbf{D_3}$ character table. The two-fold axes and σ_v planes are as indicated in Fig. 2.7.

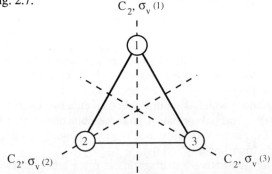

Fig. 2.7 H_3 showing symmetry operations and valence basis functions.

In order to determine which *sao's* we expect to find, we first determine the reducible representation as described above. For the three *1s ao's* of this system, the reducible representation is (3,0,0,1,1,1,3,0,0,1,1,1) which reduces to (a_1' + e'); this process is often referred to as the vector "projecting onto" (a_1' + e'). e' is a doubly degenerate representation so we expect to find one *sao* of a_1' symmetry and two of e' symmetry. We may now use any or all of the *ao's* to generate these *sao's*. In this case we find that each *ao* produces the same a_1' orbital, *i.e.*

$P^{A_I'}(s^I) = P^{A_I'}(s^2) = P^{A_I'}(s^3)$). However, different e' orbitals are produced, from which we may choose *any* two, or even any two different combinations. One possible choice of *sao's* is:

$\phi(a_1') = (s^1+s^2+s^3)/\sqrt{3}$

$\phi(e') = (2s^1-s^2-s^3)/\sqrt{6}$

$\phi(e') = (2s^2-s^3-s^1)/\sqrt{3}$

These orbitals and the qualitative energy level diagram (determined from overlap and number of nodes as discussed above) are illustrated in Fig. 2.8.

Although H_3 is not a very stable molecule, we can also use Fig. 2.8 to determine MO diagrams for molecules such as BH_3 and NH_3 by combining the B (or N) orbitals with those of Fig. 2.8. Thus, the energy level diagram for planar BH_3 may be determined by mixing the B valence orbitals[†] of a_1' (2s) and e' ($2p_x$, $2p_y$) symmetries with those H_3 *sao's* of the same symmetry to give bonding and anti-bonding combinations.

Table 2.4 The rows of D_3 character table and the transformation of *1s* orbitals (denoted **1, 2,** and **3** rather than s^1, s^2, and s^3) of the H's of H_3 under symmetry operations of D_3.

	E	C_3	C_3^2	C_2^1	C_2^2	C_2^3	σ_h	S_3	S_3^2	σ_v^1	σ_v^2	σ_v^3
A_1'	1	1	1	1	1	1	1	1	1	1	1	1
A_2'	1	1	1	-1	-1	-1	1	1	1	-1	-1	-1
E'	2	-1	-1	0	0	0	2	-1	-1	0	0	0
A_1''	1	1	1	1	1	1	-1	-1	-1	-1	-1	-1
A_2''	1	1	1	-1	-1	-1	-1	-1	-1	1	1	1
E'	2	-1	-1	0	0	0	-2	1	1	0	0	0
1	1	2	3	1	3	2	1	2	3	1	3	2
2	2	3	1	3	2	1	2	3	1	3	2	1
3	3	1	2	2	1	3	3	1	2	2	1	3

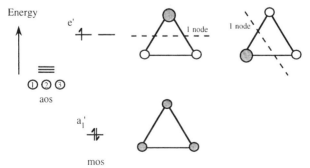

Fig. 2.8 *Mo's* and qualitative MO energy level diagram for H_3.

[†] Standard sets of character tables usually tabulate the symmetries of the orbitals of central atoms in an additional column.

Octahedral and Tetrahedral Point Groups

The final two examples we shall consider are SiF_4 and $[MnCl_6]^{4-}$. The transformation tables for these molecules are large and we could have chosen much simpler examples to illustrate the transformation properties of p and d orbitals. However, these are important geometries in inorganic chemistry and we shall use these results in Chapter 5. The transformation properties of p and d orbitals on non-central atoms are most readily identified through a two-step process: (i) the new position of the orbital is just the new position of its atom after the relevant symmetry operation has been performed; (ii) the new orientation of the orbital is the same as the new orientation of the equivalent orbital on the central atom. The transformation matrix for, say, a flourine p_x orbital on fluorine number 1 in SiF_4 is therefore obtained as a character by character product of the **x** row in Table 2.5 with the **1** row. The same principle works for determining the net character (*i.e.* the character in the reducible representation). For example, with the p orbitals on the F's of SiF_4, if j F atoms are unmoved, and k Si p orbitals remain the same but l are inverted, then the net character for the twelve F p orbitals is $j(k-l)$. We note in passing that the plethora of orbitals, symmetry operations and characters that arise for T_d and O_h symmetries necessitates rather complex notation. $(C_3^{iv})^2$ used to denote two successive C_3 rotations (240° in total) about the three-fold axis that passes through atom number 4.

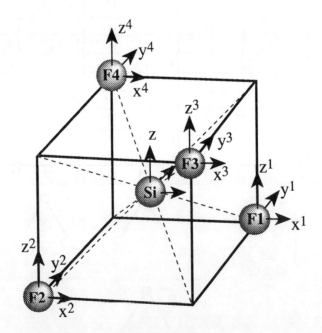

Fig. 2.9 SiF_4 inscribed into a cube to aid visualisation of the effect of symmetry operations on the component parts.

SiF_4 (T_d): Table 2.5 shows the transformation vectors of the Si p orbitals, labelled $p_x = \mathbf{x}$, *etc.* and the F 2s orbitals, labelled $s^1 = \mathbf{1}$, *etc.* (Fig. 2.9). The fate

of, say, a p_x orbital on F number 1, denoted $p_x^1 = \mathbf{x}^1$, under any symmetry operation is for it to go to the position which s^1 has adopted, and with the orientation of the p_x orbital on Si. Thus, e.g., $(C_3^{iv})^2$ takes p_x on F1 to position 3 and makes it point in the $-y$ direction: $(C_3^{iv})^2[p_x^1] = -p_y^3$.

Table 2.5 $\mathbf{T_d}$ character table and transformation table for Si p orbitals (denoted $p_x = \mathbf{x}$, etc.) and F s orbitals (denoted $s^1 = \mathbf{1}$, etc.) of SiF$_4$. C_3^i is a three-fold rotation about the axis containing atom F1; C_3^2 is two operations of the C_3 directly preceding it in the table; C_2^x is a two-fold rotation about the x-axis; S_4^x is a four-fold improper rotation about x; S_4^3 is three operations of the S_4 directly preceding it; $\sigma^{i,ii}$ is a reflection plane containing atoms F1 and F2.

	E	C_3^i	C_3^2	C_3^{ii}	C_3^2	C_3^{iii}	C_3^2	C_3^{iv}	C_3^2	C_2^x	C_2^y	C_2^z
A_1	1	1	1	1	1	1	1	1	1	1	1	1
A_2	1	1	1	1	1	1	1	1	1	1	1	1
E	2	-1	-1	-1	-1	-1	-1	-1	-1	2	2	2
T_1	3	0	0	0	0	0	0	0	0	-1	-1	-1
T_2	3	0	0	0	0	0	0	0	0	-1	-1	-1
x	x	-z	y	z	y	z	-y	-z	-y	x	-x	-x
y	y	-x	-z	x	z	-x	-z	-x	z	-y	y	-y
z	z	-y	-x	y	x	-y	x	y	-x	-z	-z	z
1	1	1	1	3	4	4	2	2	3	3	4	2
2	2	4	3	2	2	1	4	3	1	4	3	1
3	3	2	4	4	1	3	3	1	2	1	2	4
4	4	3	2	1	3	2	1	4	4	2	1	3

	S_4^x	S_4^3	S_4^y	S_4^3	S_4^z	S_4^3	$\sigma^{i,ii}$	$\sigma^{i,iii}$	$\sigma^{i,iv}$	$\sigma^{ii,iii}$	$\sigma^{ii,iv}$	$\sigma^{iii,iv}$
A_1	1	1	1	1	1	1	1	1	1	1	1	1
A_2	-1	-1	-1	-1	-1	-1	-1	-1	-1	-1	-1	-1
E	0	0	0	0	0	0	0	0	0	0	0	0
T_1	1	1	1	1	1	1	-1	-1	-1	-1	-1	-1
T_2	-1	-1	-1	-1	-1	-1	1	1	1	1	1	1
x	-x	-x	z	-z	-y	y	y	x	-z	z	x	-y
y	-z	z	-y	-y	x	-x	x	-z	y	y	z	-x
z	y	-y	-x	x	-z	-z	z	-y	-x	x	y	z
1	4	2	2	3	4	3	1	1	1	4	3	2
2	1	3	4	1	3	4	2	4	3	2	2	1
3	2	4	1	4	1	2	4	3	2	3	1	3
4	3	1	3	2	2	1	3	2	4	1	4	4

Following the projection procedures of the preceding two sections, we find that the s orbitals on the F atoms give the reducible representation $(4, 8*1, 9*0, 6*2)^\dagger$, which projects out to give $\Gamma = a_1 + t_2$, and so we need to determine one sao of a_1 symmetry and one set of three degerate orbitals of t_2 symmetry; these may be described as

$$\phi(a_1) = (s^1+s^2+s^3+s^4)/2,$$
$$\phi_a(t_2) = (3s^1-s^2-s^3-s^4)/(2\sqrt{3}),$$
$$\phi_b(t_2) = (-s^1+3s^2-s^3-s^4)/(2\sqrt{3}),$$
$$\phi_c(t_2) = (-s^1-s^2+3s^3-s^4)/(2\sqrt{3})$$

$\phi(a_1)$ has no nodes and therefore is lower in energy than the three degenerate $\phi(t_2)$ orbitals. Any three combinations of the t_2 orbitals given above are equally acceptable, as was the case for the two e' orbitals of H_3. The *mo's* may now be generated by considering overlap between these *sao's* and the Si *ao's* of the same symmetry.

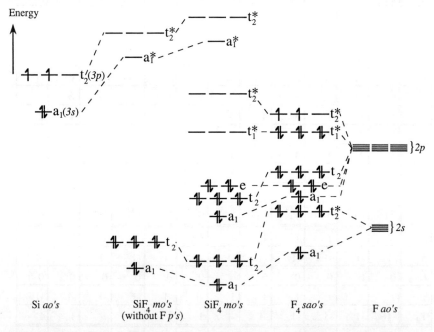

Fig. 2.10 SiF$_4$ qualitative MO energy level diagram. Symmetry labels of orbitals are shown. Dotted lines connect orbitals to their major component parts. * indicates antibonding.

There are also twelve F p orbitals, which will provide additional *sao's* that must be included if we wish to have a good description of the bonding in SiF$_4$. In this case it is certainly helpful to know where we are aiming. So, we begin by determining the vector of the number of p orbitals that remain unchanged under

---

$\dagger$ The notation $n*m$ is an abbreviation for n repetitions of the character m, thus $6*2 \equiv 2,2,2,2,2,2$.

each symmetry operation. That vector is $\Gamma = (12, 8*0, 3*0, 6*0, 6*2)$ giving twenty-four components. The scalar product of this vector with the rows of the character table given in Table 2.5, leads to $\Gamma = a_1+e+t_1+2t_2$. The symmetry adapted orbitals are given below. Only one of any degenerate pair or triplet is given. The others may be obtained from cyclic permutation of the atom labels 1,2,3,4 in the *sao's*, as was done above for the F_4 *sao's* from 2s orbitals. The first t_2 set result from projecting from $p_x^{\,1}$ and the second from $p_y^{\,1}$.

$\phi(a_1) = (p_x^{\,1}+p_y^{\,1}-p_z^{\,1}-p_x^{\,2}-p_y^{\,2}-p_z^{\,2}+p_z^{\,3}-p_y^{\,3}+p_z^{\,3}-p_x^{\,4}+p_y^{\,4}+p_z^{\,4})/(2\sqrt{3})$

$\phi(e) = (2p_x^{\,1}-p_y^{\,1}+p_z^{\,1}-2p_x^{\,2}+p_y^{\,2}+p_z^{\,2}+2p_x^{\,3}+p_y^{\,3}-p_z^{\,3}-2p_x^{\,4}-p_y^{\,4}-p_z^{\,4})/(2\sqrt{6})$

$\phi(t_1) = (2p_x^{\,1}-p_y^{\,1}+p_z^{\,1}+p_y^{\,2}+p_z^{\,2}-2p_x^{\,3}+p_y^{\,3}-p_z^{\,3}-p_y^{\,4}-p_z^{\,4})/4$

$\phi'(t_2) = (4p_x^{\,1}+p_y^{\,1}-p_z^{\,1}+2p_x^{\,2}-p_y^{\,2}-p_z^{\,2}-p_y^{\,3}+p_z^{\,3}+2p_x^{\,4}+p_y^{\,4}+p_z^{\,4})/(4\sqrt{2})$

$\phi''(t_2) = (p_x^{\,1}+4p_y^{\,1}-p_z^{\,1}-p_x^{\,3}+2p_y^{\,2}-p_z^{\,2}+p_x^{\,3}+2p_y^{\,3}+p_z^{\,3}-p_x^{\,4}+p_z^{\,4})/(4\sqrt{2})$

The two t_2 sets of orbitals thus produced are in fact degenerate as the basis functions are all related by symmetry. It is not obvious whether these t_2 orbitals are bonding, non-bonding or antibonding, but the actual *mo's* would be combinations of all the t_2 *sao's*. An energy level diagram for these ligand orbitals is illustrated in Fig. 2.10. The ordering of the levels might change depending on the degree of overlap. It should be noted that determining nodes for orbitals containing *p* orbitals must not be done simply by counting the number of minus signs in the expression for the *sao* as *p* orbitals each have both positive and negative lobes.

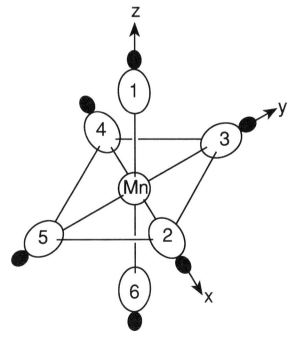

Fig. 2.11 $[MnCl_6]^{4-}$ showing ligand σ bonding orbitals.

Table 2.6. O_h character table and transformation table for Mn p orbitals (denoted p_x = **x**, *etc.*) and Cl σ orbitals (denoted σ^1 = **1**, *etc.*) of $[MnCl_6]^{4-}$ C_4^z is a four-fold rotation about the z-axis; $C_4^2 = C_2$ is two operations of the C_4 directly preceding it in the table; C_3^{123} is a three-fold rotation about the axis that rotates **1** to **2**, **2** to **3**, **3** to **1**; *etc.* $C_2^{i,ii}$ is a two-fold rotation about the axis through Cl*1* and Cl2.

	E	C_3^{123}	C_3^2	C_3^{152}	C_3^2	C_3^{134}	C_3^2	C_3^{145}	C_3^2	C_4^x	C_4^2	C_4^3
A_1	1	1	1	1	1	1	1	1	1	1	1	1
A_2	1	1	1	1	1	1	1	1	1	-1	1	-1
E	2	-1	-1	-1	-1	-1	-1	-1	-1	0	2	0
T_1	3	0	0	0	0	0	0	0	0	1	-1	1
T_2	3	0	0	0	0	0	0	0	0	-1	-1	-1
x	x	y	z	z	-y	-z	-y	y	-z	x	x	x
y	y	z	x	-x	-z	-x	z	-z	x	z	-y	-z
z	z	x	y	-y	x	y	-x	-x	-y	-y	-z	y
1	1	2	3	5	2	3	4	4	5	5	6	3
2	2	3	1	1	5	6	5	3	6	2	2	2
3	3	1	2	4	6	4	1	6	2	1	5	6
4	4	5	6	6	3	1	3	5	1	4	4	4
5	5	6	4	2	1	2	6	1	4	6	3	1
6	6	4	5	3	4	5	2	2	3	3	1	5

	C_4^y	C_4^2	C_4^3	C_4^z	C_4^2	C_4^3	$C_2^{i,ii}$	$C_2^{i,iii}$	$C_2^{i,iv}$	$C_2^{i,v}$	$C_2^{ii,iii}$	$C_2^{iii,iv}$
A_1	1	1	1	1	1	1	1	1	1	1	1	1
A_2	-1	1	-1	-1	1	-1	-1	-1	-1	-1	-1	-1
E	0	2	0	0	2	0	0	0	0	0	0	0
T_1	1	-1	1	1	-1	1	-1	-1	-1	-1	-1	-1
T_2	-1	-1	-1	-1	-1	-1	1	1	1	1	1	1
x	-z	-x	z	y	-x	-y	z	-x	-z	-x	y	-y
y	y	y	y	-x	-y	x	-y	z	-y	-z	x	-x
z	x	-z	-x	x	z	z	x	y	-x	-y	-z	-z
1	2	6	4	1	1	1	2	3	4	5	6	6
2	6	4	1	3	4	5	1	4	6	4	3	5
3	3	3	3	4	5	2	5	1	5	6	2	4
4	1	2	6	5	2	3	6	2	1	2	5	3
5	5	5	5	2	3	4	3	6	3	1	4	2
6	4	1	2	6	6	6	4	5	2	3	1	1

[MnCl$_6$]$^{4-}$ (O$_h$): We may describe [MnCl$_6$]$^{4-}$ as follows (Fig. 2.11). The Cl's are positioned at $\pm x$, $\pm y$, and $\pm z$, each with one orbital (presumably a p orbital or sp hybrid) predisposed for σ bonding and two orbitals predisposed for π bonding with the Mn. While these are not actually σ and π orbitals in the Cl atom, they will contribute to σ and π *mo's* on bonding and we shall call them σ and π accordingly; this has the added advantage of stressing that their orientations are not the same as the p_x, p_y, and p_z orbitals on the central Mn and avoids the need to specify a hybridisation scheme. We denote the Cl σ bonding orbitals as: $\sigma^1 = 1$, $\sigma^2 = 2$, $\sigma^3 = 3$, $\sigma^4 = 4$, $\sigma^5 = 5$, and $\sigma^6 = 6$. For Cl1, located at $+z$, we take its two π bonding orbitals to be parallel to $+x$ and $+y$; similarly for Cl4 located at $-x$, we take the two π bonding orbitals to be parallel to $-y$ and $-z$. The Mn is located at the origin and has p orbitals denoted $p_x = \mathbf{x}$, $p_y = \mathbf{y}$, and $p_z = \mathbf{z}$. As with F in SiF$_4$, the transformation properties of say a π_x orbital on Cl number 1, π_x^1, under any symmetry operation are identified from combining the position to which σ^1 is taken with the new orientation adopted by p_x on Mn. The symmetry properties of the Mn orbitals are summarised in standard character tables since Mn lies at the origin (see above).

As we did with SiF$_4$, we begin by finding the *sao's* for the ligands, except that this time we must use **O$_h$** symmetry. In fact, because **O$_h$** = {E,i} × **O**, the transformation table (Table 2.6) is for **O**. To complete the determination of **O$_h$** *sao's* we then operate on the **O** *sao's* with the projection operator (E+ξi)/2, where ξ = ±1. Functions that are symmetric to inversion project out under (E+i)/2, but vanish under (E-i)/2, and are labelled with the subscript "g" (standing for *gerade* meaning even); species that are anti-symmetric are obtained from (E - i)/2, and labelled with the subscript "u" (standing for *ungerade* meaning odd). To perform this projection note that x, y, and z go to $-x$, $-y$, $-z$ respectively under i, and the pairs (1↔6), (2↔4), and (3↔5) exchange.

Now, the number of basis functions of the σ-bonding set {σ^1, σ^2, σ^3, σ^4, σ^5, σ^6} that are preserved under the symmetry operations of **O** is summarised by the reducible representation Γ = (6, 8*0, 6*2, 3*2, 6*0)[†]. Upon projecting this vector with the rows of the character table we find the ligand σ orbitals have the symmetries: Γ = a_{1g}+e_g+t_{1u}. The *sao's* (again listing only one from each degenerate set) are as follows:

$\phi_\sigma(a_{1g}) = (\sigma^1+\sigma^2+\sigma^3+\sigma^4+\sigma^5+\sigma^6)/(\sqrt{6})$ [strongly bonding]

$\phi_\sigma(e_g) = (2\sigma^1-\sigma^2-\sigma^3-\sigma^4-\sigma^5+2\sigma^6)/(2\sqrt{3})$ [bonding]

$\phi_\sigma(t_{1u}) = (\sigma^1-\sigma^6)/\sqrt{2}$ [antibonding]

We now turn to the twelve Cl π *ao's*. The number of basis functions that are preserved under the symmetry operations of **O** is summarised by the vector (12, 8*0, 6*0, 3*-4, 6*0) using the above notation. Thus, we should find twelve ligand π *mo's* and they will have symmetries t_{1g}+t_{2g}+t_{1u}+t_{2u}. The member of each set resulting from the projection of π_x^1 is:

$\phi_\pi(t_{1g}) = (\pi_x^1-\pi_z^2+\pi_z^4-\pi_x^6)/2$ [antibonding]

[†] The C_4 and C_4^3 operations have identical characters, and those of C_4^2 are different. The 6*2 entry therefore corresponds to C_4 together with C_4^3 and 3*2 corresponds to C_4^2.

$\phi_\pi(t_{1u}) = (\pi_x^1 + \pi_x^3 + \pi_x^5 + \pi_x^6)/2$ [bonding]

$\phi_\pi(t_{2g}) = (\pi_x^1 + \pi_z^2 - \pi_z^4 - \pi_x^6)/2$ [bonding]

$\phi_\pi(t_{2u}) = (\pi_x^1 - \pi_x^3 - \pi_x^5 + \pi_x^6)/2$ [antibonding]

Having determined the Cl_6 *sao's* we are now ready to allow the ligands to interact with Mn to make $[MnCl_6]^{4-}$. Isolated Mn has spherical symmetry, but in the presence of the ligands it has O_h symmetry. The symmetries of its *ao's* are, in increasing order of energy : t_{2g} (from *3d ao's*); e_g (from *3d ao's*); a_{1g} (from *4s ao's)*; and t_{1u} (from *4p ao's*). The MO energy level diagram for $[MnCl_6]^{4-}$ results from taking bonding and anti-bonding combinations of these orbitals and the Cl_6 *sao's* with the same symmetry. So $\phi_\sigma(a_{1g})$ and $4s(a_{1g})$ mix; $\phi_\sigma(e_g)$ and $3d(e_g)$ mix; $\phi_\sigma(t_{1u})$, $\phi_\pi(t_{1u})$ and $4p(t_{1u})$ mix (though $\phi_\sigma(t_{1u})$ and $\phi_\pi(t_{1u})$ are orthogonal to each other); and $\phi_\pi(t_{2g})$ mixes with $3d(t_{2g})$. The strength of the Mn/Cl interactions is much greater than that of the Cl/Cl interactions as the overlap of orbitals is much greater. The various diagrams of Fig. 2.12 summarise the energy level diagrams that might be expected if different interactions are neglected. The order of the orbitals may vary with the strengths of the interactions.

Final Comment

The entries in the transformation tables above have been simple exchanges or inversions. However, C_n, $n \neq 3, 4, \ldots$ rotations often transform degenerate orbtials into linear combinations of one another. Thus, for example, when a B is placed in the centre of H_3 to make BH_3, its p_x orbital is transformed by C_3 into

$$\{-\frac{1}{2}p_x + \frac{\sqrt{3}}{2}p_y\}$$

Similar operations performed on *p* orbitals of the H's would rotate them into the same mixtures of p_x and p_y, but on the atom to which the same rotation took the corresponding *1s* orbital. Thus the transformation properties of *p* and *d* orbitals on non-central atoms may be found in a manner analogous to the way we treated F and Cl *p* and π orbitals above: determine (i) the position to which a rotation takes the atom, and (ii) ascribe to it the *combination* of orbitals to which the same orbital on the central atom would be transformed. Reducible representations may be found by adding the fractions of each orbital that remains after the operation has been performed.

2.2.2 Vibrations

All geometry changes require atoms to move and hence require vibrations of the molecule. One very useful description of these vibrations is in terms of the normal modes, since they uncouple both potential and kinetic energy to a good approximation; in other words, any vibrational energy in one normal mode does not does not affect the other vibrational modes, so that they act independently of one another. Since potential energy is $1/2kq^2$, where *q* represents the extension along the normal mode, the square of each normal mode is an observable and so must be totally symmetric. Thus, just like the wavefunctions and *mo's* discussed above, normal modes must transform according to rows of the character table. So if we determine symmetry adapted vibrations we are well on the way to determining normal modes. As with *mo's* we must have a basis set for determining symmetry adapted vibrations. This requires three coordinates per atom (*cf.* §1.2). Usually, the easiest set for ML_n systems is the set of $\{x, y, z\}$ displacements of M, and for each L, the displacement along the M-L bond, and two displacements perpendicular to the bond.

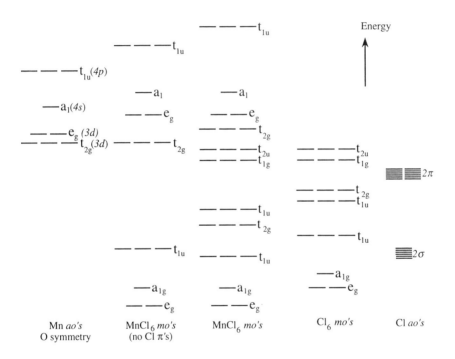

Fig. 2.12 $[MnCl_6]^{4-}$ qualitative MO energy level diagram from Mn valence orbitals and Cl bond-oriented valence orbitals. Symmetry labels of orbitals are shown.

Before launching on the task of determining a symmetry transformation table for these coordinates, we should check whether we might already have done the hard work. The answer is that we have for $[MnCl_6]^{4-}$ since the basis sets of orbitals we used for determining *sao's* for $[MnCl_6]^{4-}$ included p_x, p_y, and p_z on Mn and one Cl σ bond orbital along, and two π orbitals perpendicular to, each Mn-Cl bond. These transform in the same way as the coordinates for the atom displacements, so symmetry adapted vibrations may be read directly from the corresponding *sao's* determined above.

2.2.3 Symmetries of Wavefunctions

So far we have focused on the symmetries of individual *mo's* or vibrations. Usually, more than one *mo* or vibration is occupied and so the symmetry of the whole electronic and vibrational wavefunction is the product of these occupied *mo's* and vibrations. Since it is the square of the *total* wavefunction that is the actual observable we are usually interested in the symmetry of the wavefunction as a whole.

With *mo's*, if all the orbitals of a given energy are fully occupied, then the electron distribution from those orbitals is totally symmetric. The symmetry of the wavefunction is then totally symmetric unless there are partially occupied sets of orbitals. The partially occupied orbitals typically have electron configurations such as $(e')^1$ for H_3 or $(t_{1g})^1(t_{2u})^1$ for an O_h system. (The same products occur if the system is in the first excited state of an e' vibration or has both a t_{1g} and t_{2u} vibration singly excited.) The former situation is simple: the symmetry of the wavefunction is E' (where we use capital letters to denote symmetries of wavefunctions). The O_h example is more complicated. The answer is outlined here because we shall use the language in Chapter 5. If a more detailed discussion is required, references [1-6] should help.

If we simply wish to determine all the states that can arise from the electron configuration $(t_{1g})^1(t_{2u})^1$, we determine $\Gamma(t_{1g}) \times \Gamma(t_{2u})$. This may be done by taking the t_{1g} and t_{2u} rows of the character table and multiplying the numbers for each operation and using the answer to create a new row or vector: (9,8*0, 6*-1, 3*1, 6*-1, -9, 8*0, 3*-1, 6*1, 6*-1) with the order of operations (E, 8*C_3, 6*C_4, 3*C_4^2, 6*C_2, i, 8*S_6, 3*σ_h, 6*S_4, 6*σ_d). This is called the *direct product*. The dot product of this vector with the rows of the character table divided by h (48 in this case) will yield the symmetries that can arise. Alternatively, the direct product tables that are often associated with character tables[8] give the answer directly. The result in this case is a reducible representation with nine spatial components that reduces to: $A_{2u} + E_u + T_{1u} + T_{2u}$. The states may be either singlets (net electron spin of zero) or triplets (net electron spin of one) since we did not specify electron spin.

For two electrons in, say, the three t_{2g} orbitals, d_{xy}, d_{xz}, and d_{yz}, the situation is more restricted since the Pauli exclusion principle precludes double occupancy of any one orbital by two electrons of the same spin. The result of this is that instead of each spatial state being able to be both a singlet and a triplet, it is now either a singlet or a triplet. As previously we begin by determining $\Gamma(t_{2g}) \times \Gamma(t_{2g})$ = $A_{1g} + E_g + T_{1g} + T_{2g}$. We then need to determine a spin multiplicity (indicated as left hand side superscripts) for each state. There are fifteen ways of assigning two electrons to the three t_{2g} orbitals that are consistent with the Pauli exclusion principle. Three arrangements have spin-paired electrons in the same orbital (so $s = 0$ and spin multiplicity, $2s+1 = 1$). The functional forms of these arrangements are x^2y^2 *etc.* and they do not belong to any one row of the O_h character table. However, their $^1A_{1g}$ and ^1E_g projections resemble the *sao's* of H_3

(see above). The remaining twelve arrangements must be a singlet T and a triplet T state with respectively three-fold and nine-fold total degeneracy. Direct product tables[8] always give the antisymmetrised product of a degenerate representation with itself. In this case it is T_{1g}, so the required antisymmetry of the T_{1g} state is accounted for by the spatial part of the function and it is coupled with a symmetric (triplet) spin function. We therefore conclude that $\Gamma(t_{2g}) \times \Gamma(t_{2g}) = {}^1A_{1g} + {}^1E_g + {}^3T_{1g} + {}^1T_{2g}$. By Hund's rules the ground state has maximum spin multiplicity and so is ${}^3T_{1g}$, the others correspond to excited states.

Much the same discussion holds for vibrations except that, unlike the situation for electrons in orbitals, the number of quanta of any vibration is not restricted to two. It is interesting to note that the $v = 2, 4, 6, ...$ states of any non-degenerate vibration are totally symmetric since they transform as squares, or squares of squares *etc.* of a row of the character table whose entries are all ±1.

2.3 ML_n Geometries and their Interconversion

So far in this chapter we have examined the possible molecular symmetries a molecule may adopt, and also seen how much we can determine about *mo's*, vibrations, and wavefunctions using just symmetry. In this section we shall see how the different molecular symmetries and geometries for a given ML_n are related. In §2.3.1 we shall develop a notation for describing geometry and the relative stabilities of different geometries; and in §2.3.2 we shall look briefly at how rearrangements between different geometries of ML_n may occur.

2.3.1 ML_n Symmetry, Geometry, and Stability

The idea of molecules being described in terms of a template with holes or vacancies (§1.1) leads to a useful terminology that emphasises the relationships between different structures. The set of basic templates we adopt are polyhedra with triangulated (or deltahedral) faces. We use $\{n,0\}$ to refer to the close-packed (or *closo*) polyhedron with n vertices and no holes. So a tetrahedral molecule is $\{4,0\}$, while pentagonal bipyramidal molecule is $\{7,0\}$. Molecules that may be described in terms of an n-vertex polyhedron with one hole (*nido* or nest) are denoted $\{n,-1\}$, and those with two holes (*arachno* or spider) as $\{n,-2\}$, *etc.* It is also sometimes convenient to indicate explicitly the relative positions of two holes, for example *trans*-$\{6,-2\}$ describes an octahedral template with two holes located *trans* to one another, *i.e.* a square planar molecule. Sometimes the label is not unique, *e.g. trans*-$\{5,-2\}$ is also $\{3,0\}$. Unless there is a reason for doing otherwise, we always choose the smallest numbers for the label.

This notation is particularly appropriate for steric-plus-electronic geometry models for ML_n including the AAIM (§1.3.2), whose starting assumption is that M-L bond strengths are independent of the orientation of the L about M. It follows from this assumption that the geometry adopted by ML_n is governed primarily by the need to achieve the optimum M-L bond distance, and then by the need to maximise the L-L attractive interactions subject to a short range repulsion (due to overlapping electron clouds). If the repulsion dominates, $\{n,0\}$ structures are observed. If the attraction comes into play, $\{n,-i\}$ polyhedra are observed, since the second factor implies a tendency to maximise the L-L "contacts".

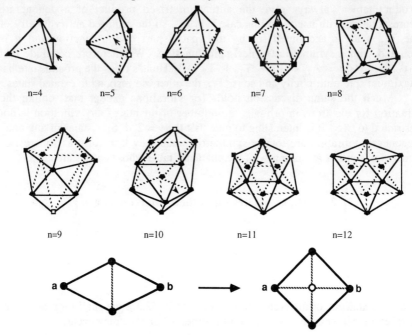

Fig. 2.13 ML_n, $n = 4 - 12$ geometries, which maximise L-L contact with little variation in M-L distance. Arrows indicate the edge which will allow most relaxation when broken. The open circles indicate the inserted vertices or interstitial holes (see text). The diamond-square-diamond relaxation mechanism is also shown. The standard representation of the capped octahedron may be obtained from the ML_8 structure shown by drawing a line between two five-fold vertices across the empty vertex.

We now turn to a thought experiment that will prove useful in understanding geometry trends in a homologous series of compounds (such as when M or L change down the periodic table), and in thinking about mechanisms for rearrangement reactions. Imagine what would happen if we steadily increased the size of M relative to L. For small M (or, equivalently, large L), the ligands will pack around M in a *closo* geometry. These polyhedra are illustrated in Fig. 2.13. Now consider successively larger M, with L remaining the same. If the molecule is forced to retain its original shape, then each L-L distance, d_{L-L}, increases, with a resultant loss of all L-L contacts. If the ligand polyhedron is then allowed to relax, it will adopt the geometry which maximises the number of L-L contacts. This means that the deltahedral polyhedron relaxes by breaking the L-L edge that allows the greatest gain in L-L interactions upon relaxation. One square face is created in the process. The arrows on the polyhedra in Fig. 2.13 indicate the edges allowing most relaxation. The relaxation process can readily be visualised when one realises that the same relaxation is required for the insertion of a new vertex (of connectivity four) along an edge of the ligand polyhedron to form a larger polyhedron. If the new L is then removed an interstitial hole is created that sits above precisely the square face that was generated in the relaxation proces described above.

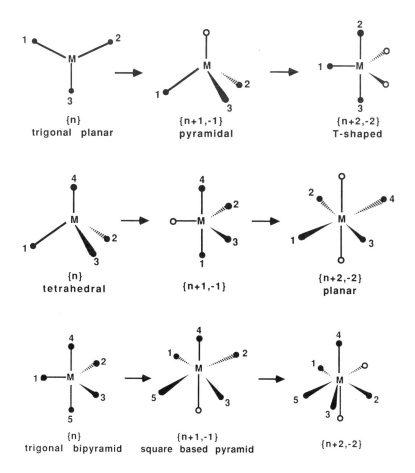

Fig. 2.14 ML$_n$ geometries, n = 3-5, showing expansion / relaxation sequence. Solid circles indicate atoms and open circles vacant sites.

For polyhedra[†] whose internal angles are less than 180°, the edge that allows most relaxation is one which is "opposite" the one, or preferably two, vertices of lowest connectivity. By connectivity we mean the number of other L it "contacts". Opposite is used in the sense that a vertex is opposite an edge in a triangular face; since each edge is part of two deltahedral faces, it will have two vertices "opposite" it, and breaking this edge will define a square face, as shown in Fig. 2.13. The resulting ligand polyhedron for ML$_n$Ω (where Ω denotes hole) defines a new *closo* polyhedron, but now with n+1 vertices. Noting that the L of

---

[†] ML$_{11}$ is an exception to this description since it is only slightly distorted from the regular icosahedron.

lowest connectivity are always the furthest from M, we see that this relaxation process enables a shortening of the longest M-L bond distances.

For even larger M, the geometry determination process may be viewed as the sequential cleavage of polyhedral edges following the sequence shown in Fig. 2.13. The choice of second, third *etc.* cleavage points relative to those already present must be the one that allows greatest geometry relaxation and so a maximum gain in the number of L-L contacts. A similar sequence is appropriate if for some reason (such as a reaction occurring) the M-L bond lengths are increased.

The expansion / relaxation sequence for ML_4 may be described as follows: one edge of the *closo*-tetrahedron, {4,0}, is broken to form the butterfly *nido*-trigonal bipyramid, {5,-1}, and a second edge broken to form the square planar *arachno*-octahedron, *trans*-{6,-2}, (Fig. 2.14c). Further relaxation may be viewed as *via* {7,-3} resulting in a pyramidal shape that would be further stabilised by relaxation to a square pyramid with equal d_{M-L}. Analogous relaxation processes are also illustrated in Fig. 2.14 for {3,0} and {5,0}. It should be noted that when the number of holes is large relative to n, the uniqueness and accuracy of the polyhedron plus hole description is lost and ceases to be helpful.

The octahedron {6,0} and the trigonal prism {9,-3} are both common six-coordinate geometries. The "expansion" sequence for ML_6 is illustrated in Fig. 2.15. For seven coordination, the *closo* polyhedron {7,0} is the pentagonal bipyramid, {8,-1} corresponds to the capped octahedron, and {9,-2} the monocapped trigonal prism. Similarly, for eight coordination we would expect the first three members of the series for ML_8 to correspond to the dodecahedron {8,0}, the bicapped trigonal prism {9,-1}, and the square antiprism {10,-2}. The cube, which is {14,-6}, may be appropriate for ML_8 with very small ligands but has not yet been observed.

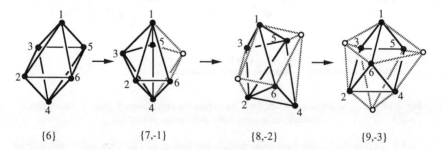

Fig. 2.15 ML_6 geometries; notation as for Fig. 2.14.

2.3.2 Stereochemical Changes

The subject of this book is molecular geometry rather than chemical reactions; however, one of the factors controlling reactivity is the geometry of reactant, product, and potential transition states or intermediates. In fact, for unimolecular isomerization reactions geometry is usually an over-riding factor. It is therefore appropriate that we consider, at least briefly, stereochemical changes between different isomers. Isomerisations of inorganic coordination and cluster compounds have become one of the most widely studied phenomena in inorganic chemistry and although the mechanisms of some are well understood, *e.g.* rearrangement processes in five- or six-coordination complexes, no simple coherent approach to all coordination numbers appears to have evolved (see references in [9-12]). When the rearrangements are viewed in terms of changes between polyhedra, a unified view does result. This chapter concludes with the outline of a formalism that lets us determine a great deal about a reaction mechanism just by knowing the symmetry of the reactant and product.

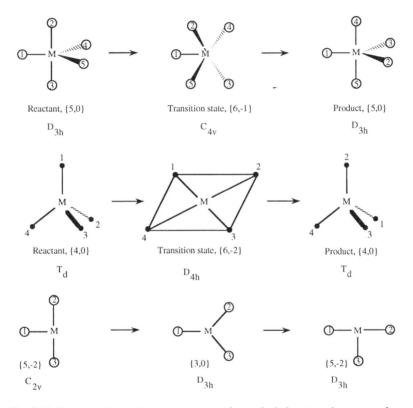

Fig. 2.16 Berry pseudo rotation; rearrangement of a tetrahedral system *via* a square planar geometry; and rearrangement of {5,-2} *via* {3,0}.

A *closo* ligand polyhedron {$n,0$} has the ligands in contact with one another, so it cannot rearrange without stretching some or all M-L distances (bond breaking is the extreme of this). So imagine the reaction of a *closo* polyhedron proceeding by first stretching the M-L bonds, followed by a relaxation of {$n,0$} to {$n+1,-1$} as discussed above. If {$n+1,-1$} can relax to a "new" {$n,0$} polyhedron - new in the sense that if the atoms could be labelled it would be apparent that different atoms are in different positions - then ML_n can rearrange by a mechanism of bond stretching and ligand relaxation *via* a transition state (or intermediate) {$n+1,-1$}. The Berry pseudo rotation[9] of trigonal bipyramidal ML_5

molecules illustrated in Fig. 2.16 is an example of such a process and we shall discuss this example further in §2.3.2. Frequently, however, {n+1,-1} cannot relax to a new {n,0}. In such a case, further M-L bond stretching and ligand rearrangement *via* {n+2,-2} may be sufficient. Rearrangement of a tetrahedral {4,0} complex *via* a square planar *trans*-{6,-2} transitions state (Fig. 2.16) is an example of this. The rearrangement of octahedral {6,0} and *tris*-chelate complexes *via* a trigonal prismatic {9,-3} transition state (see below) is an example of a reaction going *via* an {n+3,-3} transition state.

Non-*closo* polyhedra *may* rearrange *via* the same mechanisms as their *closo* analogues. However, such polyhedra are intrinsically more spacious and so one should also consider mechanisms that proceed *via less* open ligand geometries. A simple example is provided by a {5,-2} T-shaped ML_3 molecule, which can rearrange *via* {3,0} without any stretching of M-L bond lengths (Fig. 2.16). This example is discussed further in §2.3.2. Another example is provided by the square planar ML_4 *trans*-{6,-2} system which may change to a butterfly-{5,-1} ligand polyhedron without lengthening the M-L bonds. This is generally less expensive energetically than the bond stretching required to go to a larger ligand polyhedron, so if *trans*-{6,-2} can rearrange *via* {5,-1} or {4,0} rather than *via*, say, {7,-3} then it will. In this case, however, none of these options lead to a new square planar structure,[13] so the square planar compound must follow the mechanisms of its octahedral template (see below).

The utility of this approach for studying geometry and isomerization of ML_n molecules increases with the number of ligands since it enables apparently complex geometries to be systematised. For seven-coordination, there is no single-stage rearrangement mechanism available: single edge cleavage leads to {8,-1} which cannot relax to a new {7,0}. A further edge cleavage to {9,-2} (an *arachno*-tricapped trigonal prism or equivalently a monocapped trigonal prism) is required to isomerise the reactant. Hence the ligand rearrangement for a pentagonal bipyramid can be viewed as taking place by a five-stage process:

{7,0} → {8,-1} → {9,-2} → {8,-1}' → {7,0}'

{9,-2} probably forms the transition state in a concentred reaction, rather than being an intermdiate in a two step process via {8,-1} transition states.†

For eight coordination we have a situation similar to that found for five-coordination. The parent polyhedron is a dodecahedron {8,0}; single edge cleavage takes it to {9,-1} (a bicapped trigonal prism) and further extension along the same vector leads to a new dodecahedron {8,0}'. Thus its isomerization is a two stage process where the middle species is a transition state:

{8,0} → {9,-1} → {8,0}'

It is interesting to note that, {6,0}, {7,0} and {8,0} all proceed *via* the same polyhedral intermediate, {9,-i}, just as {3,0}, {4,0} and {5,0} all proceed *via* the octahedron, {6,-i}.

---

† Note the endpoint of a *stage* of a reaction may be a transition state or not even an explicit point on the reaction pathway, whereas the end of a *step* of a reaction must be (meta) stable. Thus reactant → transition state → product is one *step*, but two *stages*.

Table 2.7 M-L bond lengths required for close-packed hard sphere ligands. Bond lengths are given as $d = d_{M-L}/r$ where d_{M-L} is the M-L bond length and r the radius of the ligand.[14] Subscripts indicate the connectivity of each ligand (*i.e.* number of ligands in contact with it).

ML_3	$d_2 = 0.577$		
ML_4	$d_3 = 0.612$		
ML_5	$d_3 = 0.816$	$d_4 = 0.577$	
ML_6		$d_4 = 0.707$	
ML_7		$d_4 = 0.851$	$d_5 = 0.526$
ML_8		$d_4 = 0.930$	$d_5 = 0.677$
ML_9		$d_4 = 0.995$	$d_5 = 0.764$
ML_{10}		$d_4 = 0.995$	$d_5 = 0.764$
ML_{12}			$d_5 = 0.951$

The M-L bond stretch required for each *stage*† in a reaction scheme is a major indication of the energy required for a proposed mechanism. The stretches can be deduced from Table 2.7, where the M-L bond lengths for ML_n are determined assuming close-packed hard sphere ligands. If too much bond stretching is required in a proposed mechanism then a bond breaking mechanism will take over, as is sometimes the case for *tris*-chelate metal complexes (see below). However, in those instances where the bond stretching may be accommodated, an estimate of activation volume should be possible from the AAIM. For example, ML_6 rearranging *via* a trigonal prism {9,-3} requires the M-L bonds to stretch by 8%.[15] The corresponding volume change, or activation volume, assuming the ligands are close packed in both reactant and transition state is $[d_{M-L}{}^3(\sqrt{6}/2-2/3)]$ where d_{M-L} is the reactant M-L distance. We can compare this result with experiment for $[Cr(1,10\text{-phenanthroline})_3]^{3+}$ and $[Cr(2,2'\text{-bipyridyl})_3]^{3+}$ which are known to react *via* a twist mechanism with a trigonal prismatic transition state. Since $r = 207\,\text{pm}$[16] for these molecules the molar activation volume is predicted by the AAIM to be $3.0\,\text{cm}^3$; the experimental value is $3.3\pm0.3\,\text{cm}^3$.[17]

2.3.3 The Classical Symmetry Selection Rule Procedure

We have indicated that symmetry alone can provide a great deal of information about possible reaction mechanisms. This information may be obtained by applying the Classical Symmetry Selection Rule Procedure (CSSRP),[18-20] which makes use of symmetry changes along normal modes of vibration (§2.2.2). A strict definition of a normal mode† is a vibration that, when expressed in mass-weighted coordinates, is uncoupled to second order from all other vibrations.[21] Two key results of the procedure are that for a "well-behaved"

---

† The defintion of normal coordinate is not restricted to either stable or transition state species.

reaction path:[††] (i) the reacting motion is described by a normal mode of the system at each point; and (ii) symmetry changes occur only at the reactant, transition state, and product and these must represent high symmetry points along the reaction pathway. The CSSRP can be viewed as a procedure to determine whether there is a normal coordinate of a reactant which can take the system along a symmetry allowed reaction pathway to the product. If there is no such normal coordinate, then there can be no concerted mechanism (*i.e.* one that proceeds from reactant to transition state to product passing through no stable or meta-stable intermediates) between that reactant and product. If there is such a normal mode it describes at least the beginning of the rearrangement pathway. After an outline of the CSSRP itself, we shall give some examples of its use.

Formalism

The CSSRP is implemented by following the symmetry of a system along its reaction path. One draws the reactant (**R** of point symmetry G_R), and all products of interest (P_i of symmetry G_{P_i}), labelling their atoms. Let us once again consider the Berry pseudo-rotation illustrated in Fig. 2.16. We may write the symmetry operations of **R** and **P** using the labelling of Fig. 2.16 and the notation where $C_3^{ii}(1,5,4)$ is a three-fold rotation the tip of whose axis goes through atom 2 (denoted ii) which takes atom 1 to the position formerly occupied by 5, 5 to that formerly occupied by 4, and 4 to that formerly occupied by 1 (remember rotations are anti-clockwise by convention) - all other atoms remain unchanged. The symmetry operations of **R** are then as follows: $C_3^{ii}(1,5,4)$, $C_2^i(2,3)(4,5)$, $C_2^{iv}(2,3)(1,5)$, $C_2^v(2,3)(1,4)$, $\sigma(2,3)$, $\sigma(4,5)$, $\sigma(1,5)$, $\sigma(1,4)$. Those of **P** are $C_3^{iv}(1,2,3)$, $C_2^i(2,3)(4,5)$, $C_2^{iii}(1,2)(4,5)$, $C_2^{ii}(1,3)(4,5)$, $\sigma(2,3)$, $\sigma(4,5)$, $\sigma(1,2)$, $\sigma(1,3)$. One then determines the symmetry operations, in a labelled atom sense, that are common to each {**R**, **P**} pair. These operations form the group G_{RP}. So for our example $G_{RP} = \{C_2^i(2,3)(4,5), \sigma(2,3), \sigma(4,5)\} = C_{2v}$. These and only these symmetry operations are retained along the whole reaction path. The reacting system has symmetry G_{RP} at all points except perhaps at **R**, **P**, and **T** (the transition state, or energy maximum of the reaction pathway) where it may be higher. If no normal mode of **R** (or **P**) can reduce the system symmetry to G_{RP}, then **R** cannot convert to **P** *via a concerted* mechanism; if such a mode does exist then it provides a concerted reaction pathway. Tables of symmetry reductions possible *via* normal modes of every point group are to be found in reference [19].

There are also restrictions placed on the reaction path by the symmetry at the transition state. If reactant and product are chemically different species, the point group of the transition state, G_T, equals G_{RP}. For **R** and **P** with skeletons that are identical to or mirror images of one another, the transition state is of higher symmetry than the rest of the reaction pathway. The additional symmetry operations are determined by considering the details of how the atoms in the reactant are repositioned to make the product. The easiest way to do this is to superimpose **P** on **R** so that the symmetry elements of G_{RP} are in the same place for both, but *no other symmetry elements coincide*. Again for the Berry

---

[††] A well-behaved path is one that is a harmonic valley with respect to all motions other than the reaction coordinate, has no crossing of electronic potential energy surfaces, and has no points of inflexion along it.

pseudo-rotation, both **R** and **P** are {5,0} so if there is a concerted mechanism between them then at **T** the reacting system has higher symmetry than elsewhere along its path. To determine the symmetry of **T**, rotate **P** from its orientation in Fig. 2.16 so that atom 5 comes out of the page and atoms 2 and 3 lie in the plane of the page. All common symmetry elements of **R** and **P** now coincide but no others do. We now determine what combination of a point operation (rotation, reflection or inversion) followed by permutation of some atoms is required to take **P** back to **R**. If we denote the permutation as Π and the point operation as R, then the combination may be written $L = R\Pi$, and both L (**R**) = **P** and L (**P**) = **R**. For our example, $R = C_4^i$ superposes **P**'s skeleton upon that of **R**, but with the atoms permuted. Follow this by the permutation $\Pi = p(3,4,2,5)$ meaning atom 3 goes to where atom 4 was *etc.*, and **P** has been transformed to **R**. Now, if we apply L to **T** it must leave every atom of **T** unchanged (so it is more than a symmetry operation - it is like the identity), *i.e.* $L(\mathbf{T}) \equiv \mathbf{T}$, or $\Pi(\mathbf{T}) \equiv R^{-1}(\mathbf{T})$. This usually determines the geometry of **T**. In our example, it means that C_4^i followed by $p(3,4,2,5)$ has no net effect and **T** must be as illustrated in Fig. 2.16. It should be noted that L is not necessarily unique, but **T** always is.

The {5,-2} → {3,0} → {5,-2} rearrangement also illustrated in Fig. 2.16 provides quite a different example from that of the Berry pseudo-rotation. In this case $G_{RP} = C_S$. If we superimpose **R** and **P** with and angle of, say, θ between their two-fold axes, then a rotation of θ will take the skeleton of **P** to that of **R** and Π could then be determined. However, no **T** could be constructed for arbitrary θ. This suggests that the {3,0} geometry is in fact not a transition state but an intermediate.

The CSSRP is particularly useful for high symmetry systems, such as transition metal complexes. The assumptions underlying the CSSRP are unlikely to break down for such molecules. However, if they do, the path determined using the CSSRP will be an average of the paths that actually occur. For low symmetry systems that can be described in terms of a higher symmetry template, the mechanisms for the high symmetry template have low symmetry analogues.[13] The difference between mechanisms for a high symmetry molecule and those for a low symmetry one built on the same template is that some mechanisms that are totally impossible for high symmetry systems because they require the coalescence of atoms, may be viable in the lower symmetry case. This happens most often if the lowering of symmetry is due to atoms being removed, as is the case when we talk of square planar systems being *trans*-{6,-2}.

Applications
The Rearrangement of Polyhedra: We talked above about rearrangements of {n,0} systems being described in terms of expansion towards {n+1,-1}, {n+2,-2} *etc.* polyhedra. The CSSRP requires that if these are viable mechanisms then the beginning of the distortion from {n,0} to {n+1,-1} must be along one of the lower energy normal modes of the reactant {n,0}. Although the relative orderings of the normal modes varies as a function of the M-L bond strength, for each n the mode which resembles the {n,0} to {n+1,-1} motion most closely is in fact one of the three lowest energy vibrational normal modes.[15]

Octahedral Complexes: Fig. 2.17 illustrates the results of the CSSRP for all distinct products (in a labelled atom sense) from the symmetry allowed concerted rearrangement mechanisms of an octahedral metal complex.[14,20] The number in square brackets after each mechanism is the number of equivalent reactions for the octahedral molecule. It should be noted that two mechanisms that are equivalent (except for atom labels) for the octahedral molecule will often become different when the symmetry of the system is reduced. It is therefore important to keep track of all mechanisms of the high symmetry templates, so we can use them later.

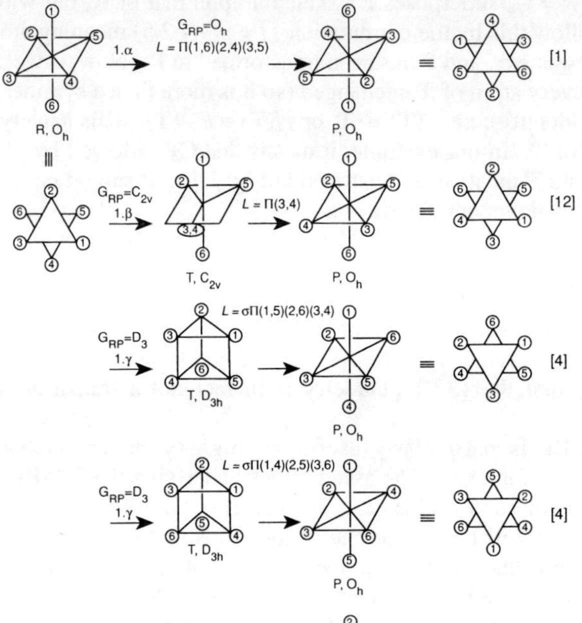

Fig. 2.17 Concerted rearrangement mechanisms for octahedral complexes. Numbers in square brackets are the number of mechanisms equivalent to the one illustrated.

Mechanisms $1.\alpha$, $1.\beta$, and $1.\delta$ are not viable for an octahedral complex as respectively, three, one, or three, pairs of atoms are coincident in **T**. (L is a pure permutation operation, and as $L(\mathbf{T}) \equiv \mathbf{T}$, this means atoms must be occupying exactly the same position.) However, these mechanisms may become operative for lower symmetry complexes and so have been included. Mechanisms $1.\gamma$

A Unified View of Stereochemistry and Stereochemical Changes 67

involve motion in opposite directions along the same T_{2u} normal coordinate of **R**; they are symmetry allowed rearrangement mechanisms which proceed *via* trigonal prismatic transition states.

Tris-Chelate Complexes: The ML_6 part of a *tris*-chelate complex $M(LL)_3$ is a good example of weak symmetry lowering from an octahedral complex. Some mechanisms which follow from relaxing the symmetry constraints on the octahedral mechanisms are illustrated in Fig. 2.18. The Greek letter label for each mechanism corresponds to the same letter of Fig. 2.17.

Products that involve *trans* ligating atoms being connected by the same chelate have been omitted, as have all products of the same handedness as **R**, since we are usually concerned about how a solution of these molecules changes from containing only one enantiomer to being a racemic mixture (equal amounts of both enantiomers, see §1.1.1). No mechanisms derived from 1.δ are illustrated as *L* is still a pure permutation for a concerted reaction.

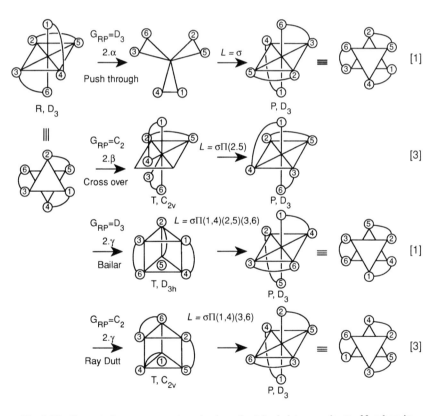

Fig. 2.18: Concerted rearrangement mechanisms for *tris*-chelate complexes. Numbers in square brackets are the number of mechanisms equivalent to the one illustrated. The labels of the mechanisms correspond to those for the analogous octahedral mechanisms.

The mechanisms of Fig. 2.18 are identical with those that result from a direct application of the CSSRP or from an exhaustive study of all possible rearrangements that one could conceive for a *tris*-chelate complex.[11] 2.α and 2.γ are both symmetry allowed concerted mechanisms. However, their transition states are very high energy structures reflecting their origin in mechanisms which were forbidden in the octahedral complexes. The common high symmetry parentage of both the Bailar and Ray Dutt twists (2.γ) is the most interesting result to come from the template approach to these reactions: the Bailar twist is a twist about the three-fold axis which is retained upon symmetry reduction from O_h to D_3, and the Ray-Dutt twist is a twist about a three-fold axis which is lost. This results from the fate of the t_{2u} normal coordinate of an octahedral **R** upon reduction of molecular symmetry to D_3: it splits into a non-degenerate a_2 component and a degenerate e component. The a_2 component gives the Bailar twist which retains D_3 symmetry along the reaction path, and the e component gives the Ray-Dutt twist. (The other direction of the t_{2u} twist falls into the group of mechanisms not illustrated as it involves a chelate becoming *trans* rather than *cis*.) The similarity of the Bailar and Ray-Dutt twists when viewed in this manner has been useful in specifying the conditions under which one would expect each to be favoured (see §5.1.3).[22]

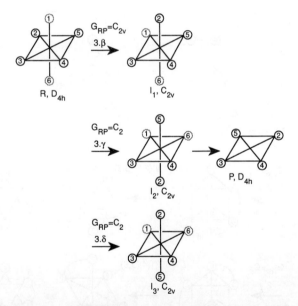

Fig. 2.19 Rearrangement mechanisms for square planar complexes. The labels of the mechanisms correspond to those for the analogous octahedral mechanisms. Ligating atom positions indicated by dotted circles are those not present.

Square Planar Complexes: The final rearrangement reaction of an octahedral parentage system that we shall examine is the *cis-trans* isomerization of square planar complexes. This is an example of what we may call strong symmetry lowering. There is only one distinct product of this reaction, and, since **R** and **P** have only one reflection plane in common, we require a normal mode that takes D_{4h} to C_s. No such vibration exists. The CSSRP therefore tells us that there is no concerted symmetry allowed mechanism. It is now that the template symmetry concept becomes so useful. 1.α-type and 1.β-type mechanisms are still not concertedly possible, though 1.β suggests a possible non-concerted mechanism to I_1 (which is probably distorted from the geometry illustrated in Fig. 2.19). However, repetition of a 1.β-type mechanism does not lead to **P**, so we also dismiss it. As in the previous examples 1.γ leads to a viable mechanism *via* I_2 (Fig. 2.18). The biggest change due to the reduction of symmetry is that 1.δ becomes an allowed mechanism (an atom and a missing atom can coalesce). However, the effect of the mechanism is the same as 3.γ, so no new mechanism emerges in this example.

References

(1) Atkins, P. W. *Molecular Quantum Mechanics;* 2nd ed.; Oxford University press: Oxford, 1983.
(2) Cotton, F. A. *Chemical Applications of Group Theory*; John Wiley and Sons: New York, 1971.
(3) Eyring, H.; Walter, J.; Kimball, G. E. *Quantum Chemistry*; John Wiley and Sons: New York, 1944.
(4) Hochstrasser, R. M. *Molecular Aspects of Symmetry*; W.A. Benjamin Inc.: 1966.
(5) Kettle, S. F. A. *Symmetry and Structure*; John Wiley and Sons: New York, 1985.
(6) Vincent, A. *Molecular Symmetry and Group Theory*; John Wiley and Sons: London, 1977.
(7) Jahn, H. A.; Teller, E. *Proc. Roy. Soc.* **1937**, *161A*, 220-235.
(8) Atkins, P. W.; Child, M. S.; Phillips, C. S. G. *Tables for Group Theory*; Oxford University Press: **1970**.
(9) Berry, S. J. *J. Chem. Phys.* **1960**, *22*, 933.
(10) Meakin, P.; Muetterties, E. L.; Jesson, J. P. *J. Amer. Chem. Soc.* **1973**, *95*, 75.
(11) J.G. Gordon, I.; Holm, R. H. *J. Amer. Chem. Soc.* **1970**, *92*, 5319.
(12) Rodger, A.; Johnson, B. F. G. *Inorg. Chim. Acta* **1988**, *146*, 37.
(13) Rodger, A. *Inorganica Chimica Acta* **1991**, 193-200.
(14) Rodger, B. F. G. J. a. A. *Inorganic Chemistry* **1989**, *28*, 1003-1006.
(15) Johnson, B. F. G.; Rodger, A.; Colwell, S. M. *Inorganica Chimica Acta* **1994**, *219*, 187
(16) Orpen, G.; Brammer, L.; Allen, F. H.; Kennard, O.; Watson, D. G.; Taylor, R.; 1986 *J. Chem. Soc. Dalton Trans.* **1989**, S1-S83.
(17) *Inorganic High Pressure Chemistry. Kinetics and Mechanism*; Van Eldik, R., Ed.; Elsevier Science Publication Company Incorporated: The Netherlands, 1986.
(18) Rodger, A.; Schipper, P. E. *Chem. Phys.* **1986**, *107*, 329.
(19) Rodger, A.; Schipper, P. E. *J. Phys. Chem.* **1987**, *91*, 189.
(20) Rodger, A.; Schipper, P. E. *Inorg. Chem.* **1988**, *27*, 458.
(21) Born, M.; Huang, K. *Dynamical Theory of Crystal Lattices*; Oxford University Press: Oxford, 1968.
(22) Rodger, A.; Johnson, B. F. G. *Inorganic Chemistry* **1988**, *27*, 3061-3062.

CHAPTER 3

The Geometry of Molecules of Second Row Atoms

Contents

Introduction		71
3.1	Geometries of ML_n, $n = 2,3,4$	74
	ML_2	74
	ML_3	76
	ML_4	76
3.2	Carbon based chemistry	77
	3.2.1 Geometry about a C	77
	3.2.2 Stereoelectronic effects	80
	Sugars	80
	Planar zig-zags or W-effects	83
	Stereoelectronics and reactivity	84
3.3	Boranes	85
	3.3.1 Bonding schemes for boranes	86
	Topological borane bonding descriptions	87
	Wade's rules	88
	The link between topological and MO descriptions of boranes	89
	Building boranes	90

Introduction

No book on molecular geometry would be complete without the molecules made by bonding second row atoms to one another and to hydrogen. However, the converse of this is that the subject matter has been well traversed. In this chapter we are primarily concerned with situations in which the central atom is a second row element; however, our emphasis will be on the concepts underlying geometry determination, especially those that will recur in later chapters.

As noted in §1.2 the second row atoms have four valence orbitals: $2s$, $2p_x$, $2p_y$, and $2p_z$ containing up to eight valence electrons. In order to make bonds between atoms, some redistribution of the atomic electron density is required. At its extremes we can identify two different types of bonding: covalent and ionic. A pure covalent bond involves equal sharing of the valence electrons, as is the case for homonuclear diatomics such as F_2 (see Fig. 1.18-19 for MO energy level diagrams). An ionic bond is exemplified by the diatomic CsF,[†] where the Cs (almost completely) donates its outer $6s$ electron to F and the molecule is then held together by the electrostatic attraction of a positive and a negative ion.[††] In general, however, different atoms in a molecule will have different abilities to attract and hold onto electrons, and so most bonds will have some ionic and some covalent character; even CsF has a small degree of covalent character.

The concept of electronegativity is a useful label to use in this context. It describes the ability of an atom to attract electrons to itself and the difficulty with which electrons are removed from it. Various quantitative measures, including the Pauling scale, can be found tabulated in chemical data books.[1,2] The Mulliken scale, whereby it is defined to be the average of the ionisation energy (E_i) and the electron affinity (E_{ea}, the energy involved in adding an additional electron to the atom), is one of the more helpful and simple definitions:

Electronegativity = $(E_i + E_{ea})/2$

A number of anomalies, such as helium having a high electronegativity due to its large E_i, arise, however, these do not really affect its conceptual use in our study of molecular geometry.

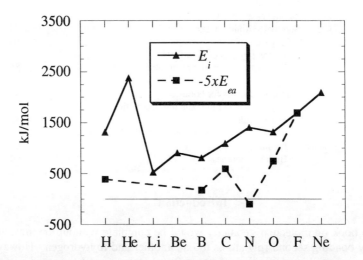

Fig. 3.1 First ionisation energies and electron affinities for first and second row atoms.

[†] LiF would be a more appropriate choice for this chapter, but CsF is much more ionic than LiF.
[††] Note that the chemist's language can be rather sloppy in this context. We often talk about the Cs as being stabilised by losing its electron. Taken at face value, this statement is patently untrue since it implies that an isolated Cs atom would expel an electron spontaneously. In fact, the statement is meant in the context of bonding, and is really saying that the energetic penalties that result from removing an electron from the Cs, are more than made up by the energy released when F gains an additional electron coupled with the electrostatic attraction between the oppositely charged ions.

As a general rule electronegativity decreases from right to left across the periodic table and from the top of a group to the bottom. The latter trend is simply because valence electrons that might be removed or added are further from the nucleus as one goes down the group and, although the nuclear charge has increased, the combined effect of (i) weaker attraction at larger distance and (ii) greater shielding of the nucleus by the core electrons, which reduces the effective nuclear charge to little more than the number of valence electrons in the neutral atom, causes electronegativity to reduce down a group. The increase from left to right is due to almost exactly opposite factors. As one proceeds across the periodic table the nuclear charge increases, yet a valence electron only partially shields electrons with the same quantum number from the nucleus, so the effective nuclear charge increases across a period; this, in turn, leads the atoms to decrease in size across a period, thus bringing the electrons closer to the nucleus and so enhancing the interaction even further. The hiccups in both E_i and E_{ea} at B (see Fig. 3.1) reflect the fact that s electrons on average penetrate closer to the nucleus and so shield other valence electrons from the nucleus slightly more effectively than do p electrons (§1.2 and Fig. 1.14); the one at N correlates with N having one electron in each p orbital, adding another requires spin pairing (which costs energy) and taking one away loses significant exchange energy (see §5.1.4).

One of the most useful guides to determining the bonding of second row atoms is the "eight-electron" rule (or more accurately the "no more than eight-electron" rule) discussed in §1.3.2. This empirical rule can be rationalised in terms of the bonds being determined mainly by the four valence orbitals (*cf.* MO ideas discussed in §1.3). So at most eight electrons are favourably accommodated about a second row atom. Since bonding interactions result in lower energy orbitals, second row atoms on the right hand side of the periodic table are described somewhat anthropomorphically as "wishing" to obtain the maximum of eight electrons in their outer shell by sharing the valence electrons of other atoms, and thus making a maximum number of bonds in the process. Atoms on the left of the periodic table, however, do not obey the octet rule for to do so would involve an excessive build up of negative charge on atoms; for molecules such as LiF_7 it would also be sterically unfavourable.[†]

The range of coordination numbers (C_N, §1.3.2) for ML_n with second row M is limited to $n = 1,2,3,4$. In contrast to many books dealing with these elements, we shall structure our discussion in §3.1 by C_N rather than by element as this relates more naturally to our concern for the geometry adopted by the molecule. Carbon based chemistry forms the subject of §3.2, with the emphasis being on geometric irregularities rather than on the norm, but with some consideration given to stereoelectronic effects. The chapter concludes with a discussion of borane geometries in §3.3. The space devoted to these last is out of proportion to the number of compounds, but is in keeping with our aim of laying a foundation for later chapters since the geometries of the cluster compounds of Chapter 6 are determined by many of the same principles.

[†] If we consider ionic crystals the situation is different with the coordination number about any one ion being typically six, but the ratio of anions and cations is such as to ensure electro-neutrality.

3.1 Geometries of ML$_n$, $n = 2,3,4$

ML$_2$

As with second row diatomics, first and second row triatomic molecules are sufficiently simple that we can usually derive a qualitative MO energy level diagram (§1.3.1, §2.2) that enables us to determine whether the geometry is linear, bent or triangular (*i.e.* with three bonds). VSEPR and AAIM (§1.3.2) may also be employed, though the limitations of both methods are most apparent for these small molecules.

The simplest triatomic molecule is the first row compound H$_3$. The valence *mo's* for this **D**$_{3h}$ molecule are shown in Fig. 2.8. Qualitative *mo's* for less symmetric triatomics usually are most easily determined by adding a third atom to a diatomic. As an example, the geometry of HCO may be determined to be non-linear by considering the changes in the MO energy level diagram for CO (Fig. 1.20) when an H is added from each of the three possible directions (linear from either end, or non-linear). The direction that most stabilises the occupied and lowest unoccupied *mo's* of CO will be favoured. The lowest occupied *mo* is included in our consideration because HCO has one more electron than CO. A non-co-linear line of approach gives favourable interaction with the antibonding π orbitals of CO, and is sufficient to ensure that HCO is bent. (The more stable molecule is HCO, rather than COH, as the interaction is greater if the H approaches from the C side: the C orbitals are closer in energy to the H *1s* than are those of O.) By way of contrast HCN is linear, as its bonding is dominated by the interaction between H *1s* and CN 5σ (the fifth σ orbital in Fig. 1.20), which is partially occupied and essentially non-bonding in the CN fragment. An MO energy level diagram for the occupied orbitals of HCN is given in Fig. 3.2.

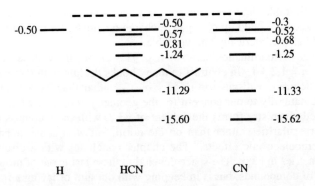

Fig. 3.2 HCN MO energy level diagram, determined as for Fig. 1.19-1.20.

HB$_2$, on the other hand, is most stable as a triangular structure resulting from the approach of H to the centre of the B-B bond. The difference from the HCN case results from a different ordering of the B_2 mo's (Fig. 1.19) and from fewer valence electrons in the whole system; in this case the perpendicular approach gives bonding interactions with all three occupied orbitals, whereas a collinear approach has most interaction with the orbitals that remain unoccupied.

It should be noted that qualitative arguments such as those given above will only be definitive when the interactions and energy changes involved are large. When the interactions are weak then the balance between the favourable and unfavourable interactions can be tricky to determine, and any firm statements should be backed up by actual calculations.

The question of the relative stabilities of linear and non-linear triatomics is more commonly addressed within MO theory by considering the orbital energy level diagram of, say, the linear molecule and deducing the changes that take place as the molecule is bent. The combined before-and-after diagram is usually referred to as a Walsh Diagram. Burdett [3] explored the applications of the Walsh Diagram approach to molecular geometry determination in some detail so we shall not pursue it here.

Table 3.1: Geometries of ML_2 compared with VSEPR, L-L attraction and L-L repulsion predictions. Angles are given in degrees. The attraction values are determined assuming hard sphere atoms of the size given in Table 1.1 (though the effect of a very electronegative atom on atomic size is noted by the inequality signs). Sources of experimental data are given in reference [4].

Molecule	Experiment	VSEPR	L-Lattraction	L-L repulsion
H-Be-H	180	180	>90	180
F-Be-F	180	180	bent	180
H-B-H	131	120	102	180
F-B-F	118	120	>111	180
H-C-H	102.4	120	<112	180
H-C-H, $^3B_1$	136	120	>117	180
H-C-N	180	180	142	180
O-C-O	180	180	>136	180
O-C-F	126	120	116	180
H-N-H	103.4	109	<129	180
H-N-C	180	120	180	180
H-N-O	108.6	120	<129	180
O-N-O	134	>120	136	180
O-N-O$^+$	180	180	165	180
O-N-O$^-$	~115.4	120	134	180
H-O-H	105.2	109	<<146	180
H-O-F	96	109	<111	180
O-O-O	116.8	109	<124	180
F-O-F	103.1	109	<99	180

In the VSEPR approach (§1.3.2) the number of bonds and lone-pairs of electrons associated with the central atom is used to establish a template on which the molecule is based. The relevant templates for second row molecules are the line, triangle or tetrahedron with idealised LML angles of 180°, 120° or 109.5° respectively (Fig. 1.22). VSEPR then recognises some distortion away from these angles due to repulsion between lone pairs and π-bonds (*i.e.* bonds made from occupancy of a π orbital) The estimates are generally quite close to experimental geometries, but there are inaccuracies; for example, F is more electronegative than H, but the FBF bond angle is greater than the HBH one. By way of contract, the AAIM optimises L-L interactions first, and although it is also generally fairly close to experiment it underrates the significance of linear geometries for ML_2 as discussed in §1.3.2. The data in Table 3.1 compare the predictions of these approaches and the L-L repulsion estimates of geometry with experiment.

ML_3

An ML_3 system can adopt two types of geometries: pyramidal or planar. Both of these may be distorted away from the most symmetric structure, although extreme distortions, such as toward T-shaped planar geometries, are rare with second row elements. VSEPR geometries are based on three templates (Figs. 1.22). In terms of the notation developed in §2.3 they are: {3,0} (trigonal planar) if there are no lone-pairs of electrons; {4,-1} (pyramidal) for one lone-pair; and {5,-2} (T-shaped) for two lone-pairs (though {5,-2} can not happen with second row elements since it requires two lone pairs and three bonds so would need valence *d*-orbitals). L-L attractions favour trigonal pyramidal geometries, whereas repulsion between the L favours {3,0}; the repulsion can arise either from electrostatic interactions or with large L. Both VSEPR and AAIM give fairly reliable predictions for the second row ML_3 molecules (see Table 3.2). For further illustrations of VSEPR see reference [5,6]. The preference of bulky non-charged ligands for a trigonal planar geometry has been illustrated clearly by Glidewell in a number of NL_3 molecules with large L.[7]

Table 3.2: Geometries of ML_3 compared with VSEPR, L-L attraction and L-L repulsion predictions determined as for Table 3.1. LML angles are given in degrees. 120° corresponds to {3,0}.

Molecule	Experiment	VSEPR	L-L Attraction	L-L repulsion
NF_3	102.2	109	<104	120
NH_3	106.5	109	<130	-
BH_3	120	120	>101	-
BF_3	120	120	>111	120
$CH_3, {}^2A_2''$	120	?<120	117	-
CO_3^{2-}	120	120	122	120

ML_4

The valence electrons required for a C_N of four means second row ML_4 are found for M on the right hand side of the periodic table. For these elements the contraction of the nucleus across the row from left to right ensures that ML_4 must be close packed {4,0} with L-L distances determined by packing and bond-length considerations. There is little scope for gross variations in shape about a "central" atom and interest in this area is restricted to the stereoelectronic subtleties of heteroatomic organic compounds (§3.2.2). Boranes (§3.3) provide an interesting exception to this since they adopt a range of ML_4 geometries that enable them to obey the eight-electron rule even though B itself only has three valence electrons.

3.2 Carbon Based Chemistry

3.2.1 Geometry About a C

From a molecular geometry point of view organic chemists have an easy life as there is little variation in the geometries about any one atom in the molecules that they study. However, the converse of this is that, synthetically, life is often more difficult as fewer options are available. The chemistry of carbon is dominated by the eight-electron rule and the formation of four bonds per carbon. Although carbenes (carbon with two bonds and one lone pair of electrons) have a wide chemistry, they are not very stable and readily react to form molecules with four bonds about the carbon.

Molecules containing carbon have identifiable two-centre single, double and triple bonds (*i.e.* approximately two, four or six electrons concentrated between a pair of nuclei). It is common to describe the bonding in terms of hybrid orbitals (*ho's*), which were described in §1.3.2 and whose radial distribution was given in Fig. 1.21, so each two-centre bond results from occupation by two electrons of a bonding combination of one *ho* on each atom of the bond. The most stable bond is always a σ bond made between two sp^n *ho's*, $n = 1, 2$ or 3, pointing towards each other along the bond axis (and hence giving good overlap). The second and third bonds (if they are present) are π bonds (§1.3.1) between *p* orbitals that were not involved in the hybridisation. The possible combinations for σ bonds are:
(i) sp^3-sp^3: a single bond, as in H_3C-CH_3
(ii) sp^3-sp^2: a single bond adjacent to a double bond, as in the H_3C-CH bond of H_3C-CH=CH_2
(iii) sp^3-sp: a single bond adjacent to a triple bond, as in the H_3C-C bond of H_3C-C≡CH
(iv) sp^2-sp^2: a double bond, as in H_2C=CH_2
(v) sp^2-sp: a double bond adjacent to a double bond, as in H_2C=C=CH_2 (note: the planes of the CH_2 groups are perpendicular)
(vi) sp-sp: a triple bond as in HC≡CH, or a double bond adjacent to two double bonds, as in the central bond of H_2C=C=C=CH_2

The templates for the carbon geometries are either tetrahedral, trigonal planar or linear (Fig. 1.22), depending on the combination of *ho's* involved, and can be worked out from the number of single, double, or triple bonds associated with each C. It should be noted that the C-C 'single' bonds in cases (i-iii) are all different, and one might reasonably expect a difference in both C-C bond energy and bond distance. In fact there is remarkably little variation (~5%), which allows C-C distances and bond energies to be considered as transferable properties for a whole variety of different organic molecules and, as a consequence, firm predictions of stability and reaction mechanisms are possible. A difference does come with molecules such a benzene (Fig. 3.3) where it is not possible to identify any one C-C bond as single or double. In that case the C-C bond length is a compromise between a single and double bond. This consistent behaviour of carbon in many different situations means that molecular modelling theories can be made successful with carbon based systems. The successes of molecular mechanics (§1.3.2) in studies of macromolecules such as DNA and proteins (see Chapter 7) illustrate this.

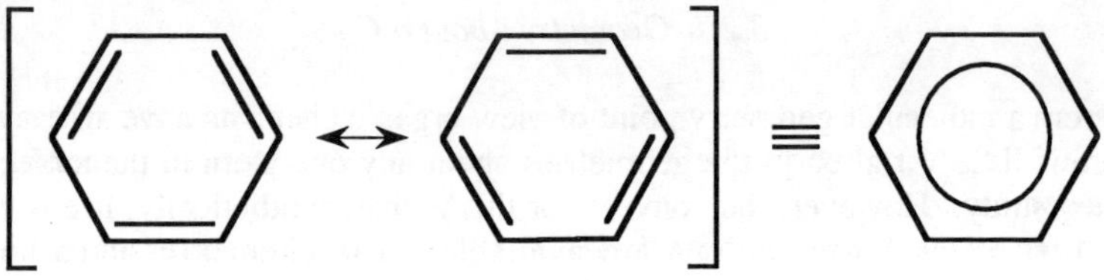

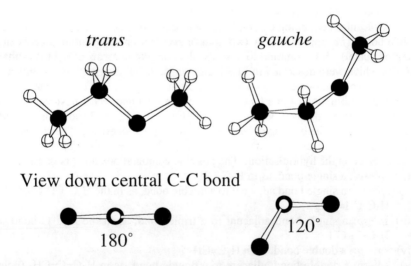

Fig. 3.3 Benzene.

Fig. 3.4 *Trans* and *gauche* conformations of butane.

Carbon also bonds to other main group atoms, and does so with the same types of bonding. The shapes of the molecules are entirely consistent both with the VSEPR model and with the packing expectations of atoms of different sizes and electronegativity that underlies the AAIM method (§1.3.2). Thus, the nature of carbon is to have entirely predictable geometries about any one atom. The difficult geometry questions for carbon systems arise from the fact that carbon readily forms catenated systems whose long chains and branches have a wide variety of geometries and isomers available to them. When chains of carbons are considered there are far more possibilities. Even in the simplest case of ethane, rotation about the C-C bond leads to different geometries as discussed in §1.1.1 and illustrated in Fig. 1.8. More complicated examples occur as the chains gets longer, and in these cases interest tends to centre on the extent to which the backbone bonds are "twisted". We define the torsion angle by looking at the two atoms that form the bond, and two more atoms that are attached to the bonded atoms. The simplest example for which such conformational changes are chemically important is butane, for which the rotation about the central bond leads to distinct *gauche* and *trans* conformers (Fig. 3.4). The *gauche* and *trans* conformations have different shapes and therefore different physical properties. In particular, the *trans* conformation is flat molecule (the arrangement of C's is planar) whereas the *gauche* conformation is more compact. Consequently, adsorption onto a surface at low coverages leads to a high percentage of *trans* molecules (so that more of the molecule can lie flat on the surface), whereas in liquids, where there is very little free space, one finds an increased percentage of *gauche* conformations.

An important application where long alkane chains is essential is in Langmuir-Blodgett films - dense and highly ordered layered structures of precise molecular thickness that can be formed on many surfaces; in these systems the presence of *gauche* bonds will give a significant disruption of the desired structure. For even larger molecules the question of conformational shape becomes even more complicated, but even more important. Perhaps the supreme example is protein folding (§7.2): much of the time enzymes will adopt a "dormant" structure in which the active site is well protected, but this means that, the molecule must undergo substantial conformational changes (*i.e.* many of the bonds must undergo large twists) in order to make the active site available.

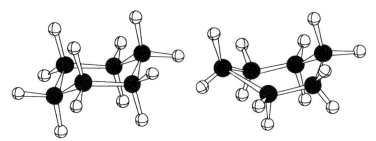

Fig. 3.5 Chair and boat conformations of cyclohexane.

A particularly interesting type of conformational behaviour arises when a chain of carbons is joined up to form a ring. In this case the constraint of forming a ring will place some strain on the molecule, and the ring will pucker in order to alleviate this strain. Any twisting about bonds in the ring must happen in a concerted manner, and involve more than one bond in substantially greater energy penalties than is involved in twisting a single bond. The classic example of this type of behaviour is the chair and boat conformations of cyclohexane (Fig. 3.5). In fact the boat conformation has not yet been observed, but is postulated as an intermediate or transition state in many reaction pathways. The situation becomes more complex if one of the C's is replaced by a heteroatom such as N or O. In this case, two different chair forms of may be identified, differing in the location of the heteroatom within the chair. Probably the most important occurrence of this is in sugars and is discussed further in §3.2.2.

Five-membered rings may be drawn to be planar with average bond angles of 108°, rather than the tetrahedral angle of 109.5°, but in practice they also pucker and become non-planar. Which atom in a hereto-atomic ring puckers has important consequences for the geometry of DNA, as discussed in §7.1. In general, the distortion will tend to bring groups attached to the ring closer together, particularly if they point out from the local plane of the ring (equatorial). Indeed, the L-L repulsion can become so important in such systems that it can dominate even over bonding interactions.[9] However, two-centre bonds and L-L interactions do not always tell the whole story, as the next section illustrates.

3.2.2 Stereoelectronic Effects

In recent years it has become apparent that organic molecular geometry is not completely specified by the sum of electronic and steric effects (§1.3.2).[10] The extra factors have been loosely described as stereoelectronic effects. In some contexts the label stereoelectronic has been used to mean something very specific, this has generated a great debate as to whether stereoelectronic effects exist and if so what their origin is. A more general definition of these stereoelectronic factors is that bond strengths are dependent upon the geometry a molecule adopts, so that there is significant coupling between steric and electronic factors. In many systems the stereoelectronic effects are so small that it is safe to ignore them completely; however, there are systems in which stereoelectronic factors will dominate over steric interactions and it is important to understand when and why this happens. A particularly simple example is 3-acetoxy-tetrahydropyran where steric effects favour the equatorial conformation, illustrated on the left of Fig. 3.6, but axial is more often observed.[10]

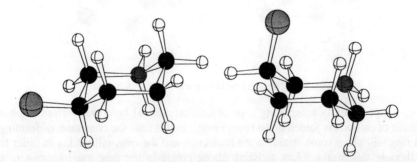

Fig. 3.6 Equatorial and axial 3-acetoxy-tetrahydropyran. O is denoted by the medium size shaded sphere, and the acetoxy group (-OCOCH$_3$) by the large shaded sphere.

Sugars

The molecules for which stereoelectronic effects have been most contentious are sugars.[11,12] Sugars are always in an equilibrium between their ring form and the open chain aldehyde (or keto in, *e.g.*, fructose). This continual opening and closing of the ring means that in addition to the two different chair conformations possible, any sugar also has two possible stereochemistries at C^1, denoted α and β and usually referred to as different *anomers* (Fig. 3.7). Thus the combination of two chair forms with two anomers gives a total of four different ring isomers, and the relative population of each geometry reflects their relative stabilities. On purely steric grounds, a maximum number of equatorial substituents is optimal. In glucose (Fig. 3.7) this has the consequence that 64% of the glucose molecules take the β form since this has all substituents in the equatorial positions. Mannose differs from glucose only in that the stereochemistry at C^2 has been inverted. However, at equilibrium the α-mannopyranose form is more stable than the β-mannopyranose (69% α versus 31% β), so for some reason the OH on C^1 adopts the sterically less favoured axial position. This behaviour, known as the anomeric effect, was first noted by Lemieux.[13]

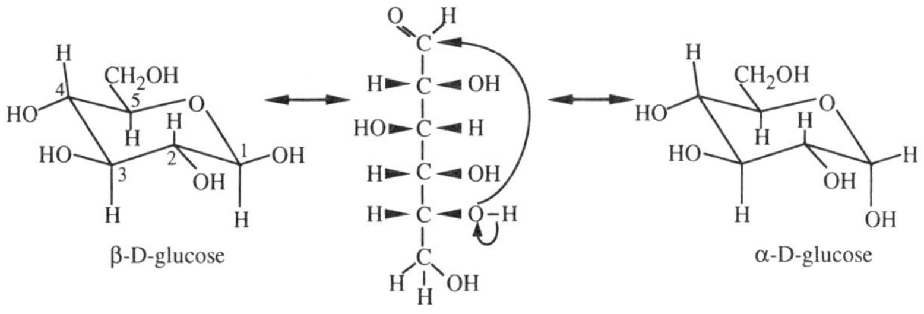

Fig. 3.7 α and β anomers of D-glucose, showing the aldehyde intermediate.

It is now accepted that there are two major contributory factors to the anomeric effect. The first is the destabilising effect due to repulsion by lone pairs on 1,3 atoms, in this case O's (Fig. 3.8). In saying this we have adopted the language of the VSEPR model, but this explanation does have its analogue in the other models as well: in MO terms it is ascribed to the bonding and anti-bonding interaction of two occupied orbitals, which results in two occupied orbitals that are net destabilised compared with the initial ones (§1.3.1). The second factor is generally agreed to be more important (but see [14]), and turns out to be a breakdown in the fragment MO arguments in which we considered all bonds to be made from overlap of localised orbitals on neighbouring atoms (or functional groups) and sharing of electrons between those groups. In fact, the nature of one bond made by an atom affects all its other bonds and the result can be a net delocalisation of the electrons even in the σ bond framework. We shall continue to discuss the anomeric effect in MO terms, but note that various polar and simple electron migration descriptions have also been given for the same phenomenon.[10]

We first consider the simpler case of O-C-X, where X is an electronegative atom and all the orbitals have been allowed to hybridise (*cf.* §1.3.2) ready for bonding. If the O is oriented so that one of its lone pairs is anti-parallel to the C-X bond (this is called "antiperiplanar"), then a π-type overlap can occur between the back lobe of the lone pair and the σ (valence σ bonding) and σ* (valence σ antibonding) orbitals of the C-X bond; this overlap is ignored if we consider the molecule as a collection of localised bonds. The interaction between the lone pair and the C-X σ* orbital is the more significant (Fig. 3.8), since, in this case, the additional interaction is bonding in character and can result in a net transfer of electron density into other parts of the molecule. The C-O bond is strengthened and the C-X bond weakened as σ* becomes part of an occupied orbital. The C character of the σ* orbital, and hence the interaction, increases with the electronegativity of X since greater overlap with the lone pair occurs.

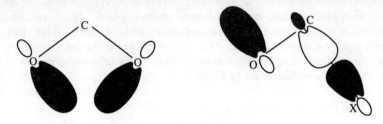

Fig. 3.8 1,3 lone pair - lone pair repulsion, and antiperiplanar orientation of O lone pair with respect to the C-X σ bond.

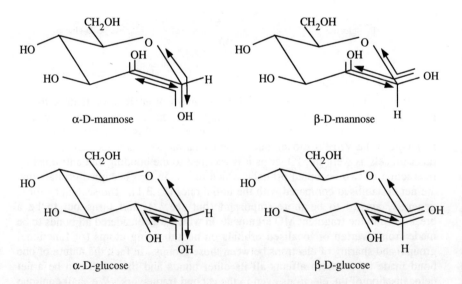

Fig. 3.9 α and β anomers of mannose and glucose indicating the favourable stereoelectronic interactions with the direction of electron transfer indicated by arrows.

In the case of sugars the O-C-X system is formed by the ether O, C^1 and the OH group; as both O's may fill either role, the O-C-O grouping has a double strength stereoelectronic effect. Furthermore, for mannose both C^1 and C^2 may be considered as electronegative atoms (as they are bonded to O), so we get two smaller contributions as illustrated in Fig. 3.9. (One must check whether it is possible to orient the orbitals so that all four effects occur simultaneously - in this case it is.) However, this does not explain why the β-anomer is more stable for mannose but not for glucose, since the same number of stereoelectronic interactions may be envisaged for glucose.

We must now ask how having an axial OH on C^2 in mannose enhances these stereoelectronic effects, or conversely, why it is reduced for glucose. As usual there is more than one factor contributing to the observed effect. Significant factors include: (i) O^2 will draw electron density away from C^2, thus increasing the interaction between O^1 and the C^1-C^2 σ* orbital; and (ii) in order for α-glucose to have its stereoelectronic interactions, either the H^1 and O^2, or the non-anti-periplanar lone pair on O^2, will clash with the O^1 antiperiplanar lone pair. A staggered arrangement losing all stereoelectronic effects is therefore preferred. These effects are small, but for mannose only 2kJ/mol energy difference is needed to ensure the experimental population difference. In fact, glucose does show some stereoelectronic stabilising of the α-anomer as steric factors alone would suggest 82% preference for β.[10]

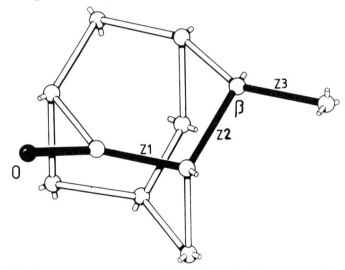

Fig. 3.10 β-equatorial adamantanones. Z1, Z2, Z3, and C=O form a planar W zig-zag.

Planar Zig-Xags or W-Effects

Molecules with a planar zig-zag (Fig. 3.10) of bonds have been noted as having particularly strong 1-4 interactions[†] in both NMR[15,16] and circular dichroism (CD) spectroscopy.[17,18] The same type of geometry dependent electronic interactions as noted for sugars operate here. Consider the carbonyl chromophore of β-equatorial adamantanones, for which CD is a convenient probe of excitations from the $n \to \pi^*$ orbitals of the CO group (see Fig. 3.11). The bonds labelled Z1, Z2 and Z3 in Fig. 3.10 form a coplanar zig-zag with the C=O of the carbonyl. Overlap of occupied and unoccupied orbitals along the zig-zag chain is illustrated in Fig. 3.12 for both the n and π^* orbitals, along with the

---

[†] An interaction between an atom and one three bonds away.

resulting distortion of orbital energy level diagrams. The net effect is a red-shift of the transition (since n increases in energy and π^* is lowered), with the magnitudes being determined by the identity of the atoms along the chain. If the conformation is changed, then the degree of overlap is decreased and the effect is reduced significantly.

Stereoelectronics and Reactivity

Stereoelectronic effects are also observed in the reactions of systems with electronegative elements that have non-bonding electrons available for stereospecific electronic interactions. Fluorine's size (C-F bonds are 1.38Å, C-H bonds are 1.08Å) enables it to be substituted for H atoms in organic molecules without any substantial steric problems. It is often assumed that the fluorine atom does nothing except prevent H abstraction reactions, but this assumption is not always appropriate. An extreme example of the influence of F on reactivity is provided by α-trifluoromethyl ketones.

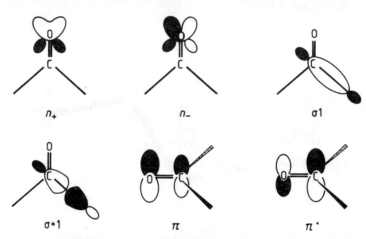

Fig. 3.11 Schematic illustration of the carbonyl valence orbitals. n_+ and n_- are the in-phase and out-of-phase couplings of the sp^2 lone pairs in accord with the C_{2v} symmetry of the carbonyl (see §2.2.1). Shading indicates relative phase.

F_3C^1-C^2O-R is much more prone to attack by a water molecule to form F_3-$C(OH)_2$-R than is H_3C^1-C^2O-R. This reactivity is not expected on the basis of traditional electronegativity or electrostatic arguments, but can be accounted for in terms of the relative stabilising effects of the interactions of the lone pairs of one (in the carbonyl) versus two (in the hydrate) oxygen atoms with the C^1-C^2 σ^* orbital. Due to the combined electron withdrawing power of the three F atoms attached to C^1, the σ^* orbital has more C^2 character than is usual in carbonyl compounds, so the stabilising interaction of a lone pair which is "antiperiplanar" to the C^1-C^2 bond is larger than in the unsubstituted case. In addition, the C^1-C^2 σ orbital has less destabilising interaction with the O lone pair(s) than in the unsubstituted molecule because the bonding orbital has less C^2 character and hence less overlap with the lone pairs. In the hydrated molecule one lone pair on each O has this stabilising interaction.

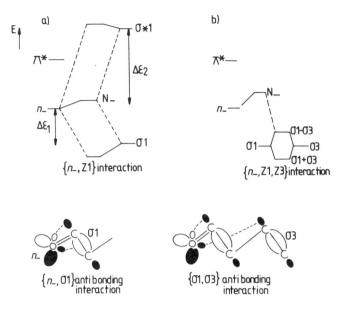

Fig. 3.12 (a) Interaction of the sigma bonding ($\sigma 1$) and sigma anti-bonding (σ^*1) orbitals of Z1 with n_-; (b) the effect of $\sigma 3$ on n_- via its coupling with $\sigma 1$.

3.3 Boranes

The difference between boron and carbon is small – one proton, one electron, and one or more neutrons – and in fact they have many similar features. $B_3N_3H_6$ forms an aromatic compound resembling benzene and carboranes differ from boranes by the substitution of C-H for BH_2. In general, however, their chemistries and molecular shapes are very different. The means adopted by B to achieve an octet of electrons involves the formation of cage-like structures with three-centre two-electron bonds, whereas for C, three-centre bonds would force too many electrons into the valence shell.

Bonding in molecules composed of B and H atoms, the boranes, is often seen as rather unusual. For example the regularity of carbon chemistry (see above) and of much main group chemistry – with clearly defined bond types of very nearly constant length and with predictable geometries – is not evident in boranes. Some of the simpler boranes are illustrated in Fig. 3.13 with some B-B and B-H bond lengths indicated. Boranes are highly reactive and difficult to prepare, initially there was hope that they might prove to be effective rocket fuels as they have positive enthalpies of formation relative to B_2 and H_2. For our purposes they are most valuable as simple templates for the molecular geometries of transition metal clusters to which we shall come in Chapter 6.

Before looking at why boranes adopt such different molecular geometries from carbon based molecules it is helpful to catalogue the shapes they do adopt. If we ignore the H's (*i.e.* just assume they are there to complete the electron and orbital overlap count) and connect each B to its nearest neighbours with a line then the molecules are all three dimensional deltahedral polyhedra; the B atoms form the vertices of the polyhedron, and the "lines" (they are not necessarily conventional bonds as we shall see below) form the edges. We shall use the notation developed in §2.3.1 for such polyhedra.

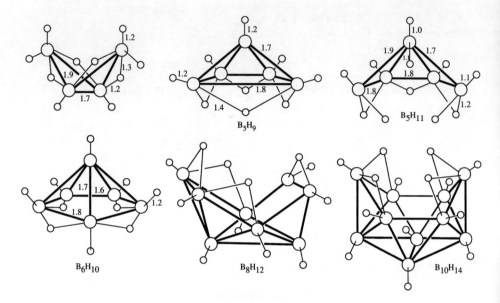

Fig. 3.13 Some boranes.

It was Wade[20] who first noted that if there were twelve or fewer boron atoms in the polyhedron then the structure could be described either as a *closo* regular polyhedron, or as a fragment of one with missing vertices. The fragments were labelled *nido, arachno, hypo* having respectively one, two or three vertices missing. In the notation developed in the previous chapter these are: $\{n,0\}$, $\{n,-1\}$, $\{n,-2\}$, and $\{n,-3\}$. $B_{13}H_{13}^{2-}$ forms a structure similar to the icosahedron but with one *extra* vertex inserted. Still larger boranes are composed of smaller ones fixed together either at a vertex or along edges. Fig. 3.14 illustrates the parent B_n polyhedra. The H's are then stuck onto the vertices of the polyhedron. They may be either terminal, so attached to only one boron and directed out from the polyhedral surface, or bridged between two (or perhaps three) B's. All B's have at least one terminal H unless they are the connecting link between two polyhedra.

3.3.1 Bonding Schemes for Boranes

A number of schemes have been developed to account for the structure and bonding in boranes. These are reviewed in reference [20]. Broadly speaking, these may be divided into those based on topological (connectivity) arguments and those based on MO theory (usually qualitatively or semi-quantitatively).

Philosophically the difference between these two types of approach lies in the fact that *mo's* are delocalised over the whole molecule, whereas topological approaches use localised bonds between nearest neighbour atoms. In the following we give a brief outline of the general ideas behind these two approaches and describe three schemes derived from them: the topological approach (TA), the extended topological approach (ETA), and Wade's rules.

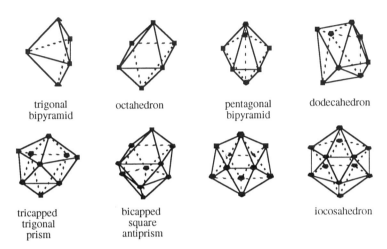

trigonal bipyramid   octahedron   pentagonal bipyramid   dodecahedron

tricapped trigonal prism   bicapped square antiprism   iocosahedron

Fig. 3.14 Borane polyhedra. The number of nearest neighbours to each vertex is indicated by the polygon at the vertex.

Topological Borane Bonding Descriptions

The idea of boranes being held together by a mixture of localised two- and three-centre two-electron bonds (*i.e.* two or three atoms held together by the electron density of two electrons) was first developed by Longuet-Higgins [21,22] and has since been exploited extensively. On the assumption that a borane is most stable if each B has a nominal count of eight valence electrons and each H has two electrons, then each B is best represented by four $2sp^3$ *ho's* and each H by a *1s* orbital. Then, a two-centre two-electron bond is the bonding interaction either of two $2sp^3$ *ho's* or one $2sp^3$ *ho* and one *1s* orbital (Fig. 3.15). A three-centre two-electron bond is either bent or triangular as illustrated in Fig. 3.15, however, no description of BBB bonds are entirely satisfactory. Calculations on various boranes and carboranes (boranes with one or more B atoms replaced by C) have electron density maps which show concentrations of electron density between neighbouring atoms in a way that supports the concept of two- and three-centre bonds.[23-25]

The TA of Lipscomb *et al.* [23-25] follows from these ideas and is based on the following assumptions.
(i) Two electrons are required to make any of: a BH bond to a terminal H; a three-centre BHB bond; a three-centre BBB bond; or a two-centre BB bond.
(ii) Each B has at least one terminal H.
(iii) All B's separated by a distance that indicates a bond are connected by one two-centre bond, or one or two three-centre bonds (BBB or BHB).

(iv) All B valence orbitals are involved in bonding.

Point (ii) implicitly limits consideration to convex fragments; however, modification to include linked polyhedra is straightforward if we write the borane as $B_{p+r}H_{p+q}{}^{c-}$ where r is the number of B's without a terminal H.[20] the arithmetic balance of valence electrons (three per B and one per H) and valence orbitals (four per B and one per H) gives constraints on the number of two- and three-centre bonds:

$$n_{BBB} = p + r - n_{BHB} - c \geq 0$$

where n_{BHB} is the number of BHB three-centre bonds *etc.* and

$$n_{BB} = n_{BHB} + r/2 - q/2 + 3c/2 \geq 0$$

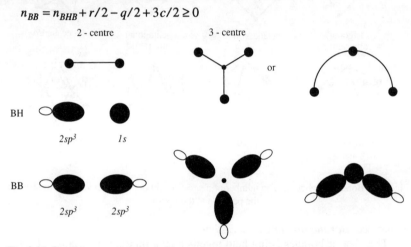

Fig. 3.15 Possible representations of two-centre two-electron bonds and three-centre two-electron bonds in boranes.

The TA can be extended (leading to the extended TA or ETA) [26] if three extra parameters are included: n_e, the number of edges in the boron polyhedron; n_d, the number of edges shared by two three-centre bonds; and n_2, the number of B's with two terminal H's. We can then write:

$$n_d = n_{BB} + 3n_{BBB} + n_{BHB} - n_e \geq 0$$
$$= 3p - n_{BHB} - 3c/2 - q/2 + 7r/2 - n_e \geq 0$$
$$n_2 = q - n_{BHB}$$

Note that $n_e = 3p-6$ for $\{p,0\}$. Examples to illustrate the use of these equations are given below.

Wade's Rules

The MO theory used for studying boranes has usually been of the simplest kind, though some very sophisticated calculations have been performed. In general terms it has been most successful when used semi-quantitatively. The success of comparatively simple calculations is in large part due to the constraints imposed by the high symmetry of the molecules. Wade's rules for *closo* boranes are the most widely used set of results from MO studies.[20]

From considering the results of MO calculations on *closo* boranes of molecular formula $B_pH_p^{2-}$ Wade deduced that a *p*-vertex polyhedron made up of B's with three orbitals available for framework bonding (since the other one is used for its bond to a terminal H) has (p+1) bonding orbitals, and so ideally 2(p+1) electrons go into framework bonding orbitals and 2(p) into B-H bonding orbitals. It is the 2(p+1) framework bonding orbitals that result in $B_pH_p^{2-}$ being observed, not B_pH_p.

Wade's observation that neutral boranes were simply fragments of *closo* polyhedra enabled him to extend these rules. The argument goes as follows. The removal of one vertex from {p,0} does not alter the symmetry characteristics of the *mo's*, therefore {p,-1} would also have (p+1) bonding *mo's*. So, *nido*-$B_{p-1}H_m$ would require (p+1) pairs of skeletal bonding electrons for maximum stability. And so on. In this context, the bridging H's are counted as contributing their electron to skeletal bonding. The rules break down as soon as any ambiguity in the fragment description of a particular borane occurs; however they are remarkably successful and have been extended to transition metal clusters.

The Link Between Topological and MO Descriptions of Boranes

Molecular orbitals are usually delocalised over the whole molecule; however, as discussed in §1.3.1, much of chemistry may be described in terms of properties of relatively small fragments of molecules. Organic chemists refer to functional groups, and spectroscopists to chromophores. Similarly, we can "build" boranes by joining together molecular fragments. In the process, orbitals from the molecular fragments (*mfo's*) will interact and may overlap to give bonding and antibonding interactions.

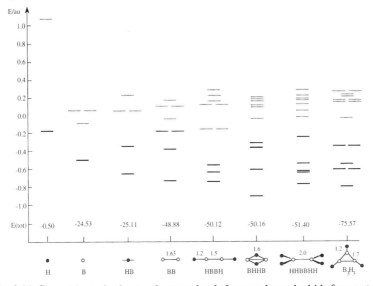

Fig. 3.16 Geometries and valence *mfo* energy levels for some boron hydride fragments. Dotted lines indicate energies of unoccupied orbitals.

Although individual *mfo's* have no reality in themselves, the changes in the orbitals of two molecular fragments A and B indicate how the wavefunction – and hence the electron density and bonding – changes upon molecule formation. The most important A and B *mfo's* will be those of comparable energy that have significant overlap. The main changes upon molecule formation are therefore due to *mo's* of AB that are occupied, but have a large component from *mfo's* that are un- or partially occupied in a constituent fragment, *i.e.* those resulting from the interaction of higher occupied *mfo's* with lower unoccupied *mfo's* (so called frontier orbitals). The various fragment energy level diagrams in Fig. 3.16 were calculated on a PC using the program MICROMOL.[27] Although these calculations are too simple to give quantitative accuracy, the relative energetics and trends are well reproduced and the results are not obscured by too much detail. Fig. 3.17 shows analogous calculations for hydrocarbon fragments, and Fig. 3.18 calculations for ethane and diborane. It is interesting to note that there is so little difference between them, but that difference is sufficient to cause the divergence of boron and carbon chemistries.

Building Boranes

Using (i) the constraints of the ETA, (ii) the fact that the order of energetic stability of bonds is BBB ≥ BHB ≥ BB ≥ BH, and (iii) the fact that for steric reasons only one terminal H may be accommodated if a B has more than two nearest neighbour B's, it is possible to deduce stable borane geometries, or conversely the number of electrons required for stability. Pentaboranes are discussed below. Additional examples may be found in reference [26].

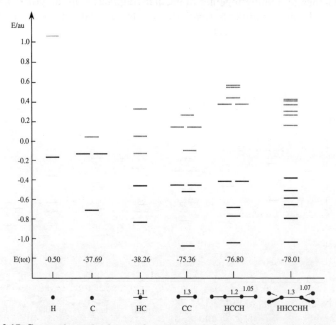

Fig. 3.17 Geometries and valence *mfo* energy levels for some hydrocarbon fragments. Dotted lines indicate energies of unoccupied orbitals.

Pentaboranes: The possible triangulated topologies for B_5 are $\{5,0\}$, $\{6,-1\}$, and *cis*-$\{7,-2\}$ (Fig. 3.19).

Consider $\{5,0\}$ with molecular formula $B_5H_{5+q}{}^{c-}$. The number of skeletal bonds may be deduced from the ETA equations given above. $n_e = 3 \times 5 - 6 = 9$; $n_{BHB} = 0$ (for steric reasons); and $n_2 = q = r = 0$ as all B have three or four nearest neighbour B's. So, there are three possible $\{5,0\}$ molecules with different values of c: (i) $c = 0$, $n_{BBB} = 5$, $n_d = 6$, $n_{BB} = 0$ with a total of ten skeletal bonds; (ii) $c = 2$, $n_{BBB} = n_{BB} = n_d = 3$ with eleven skeletal bonds; (iii) $c = 4$, $n_{BBB} = 1$, $n_d = 0$, $n_{BB} = 6$ with twelve skeletal bonds. The more bonds the molecule has the more bonding energy it has (with the order of bond stability as noted above), but the greater its charge the higher its energy. In this instance the optimum balance is given by (ii) and we observe $B_5H_5{}^{2-}$ (as predicted by Wade's rules).

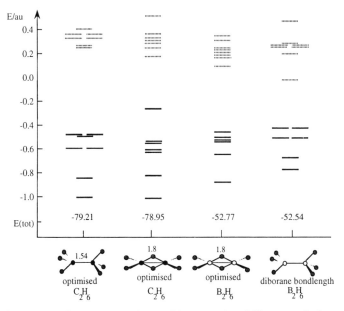

Fig. 3.18 Valence MO energy levels for possible geometries of diborane and ethane. Dotted lines indicate energies of unoccupied orbitals. (To make semi-quantitative deductions from orbital energy level diagrams it is best to compare different geometries with the same bond lengths so that the nuclear repulsion is the same.)

$\{6,-1\}$ and $\{7,-2\}$ are more open structures than $\{5,0\}$, which means that bridging H's, rather than electrons, may be used to complete the required electron count. Thus we can assume that the most stable species are neutral. $\{6,-1\}$ has $n_e = 8$, $n_2 \geq 0$, and $n_{BHB} \leq 4$. If the number of bridging H's is in fact four (the maximum that can be sterically accommodated, thus ensuring the maximum number of three-centre bonds), then $n_2 = 0$ (for steric reasons), $n_{BBB} = 1$, $n_{BB} = 2$, and $n_d = 1$ (by the ETA equations), so the molecular formula is B_5H_9. The bonds may then be assigned as in Fig. 3.19.

$\{7,-2\}$ has $n_e = 7$, $n_2 \geq 2$, and $n_{BHB} \leq 4$ (for steric reasons only one of the "front" B-B edges may accommodate a bridging H). The maximum number of bonds has $n_{BHB} = 4$ and steric considerations then imply that $n_2 = 2$, so $n_{BBB} = 1$,

$n_{BB} = 1$, and $n_d = 1$, and the molecular formula is B_5H_{11} (Fig. 3.19). A "resonance" between this geometry and its mirror image would have additional stability since little or no geometry distortion is required. This "resonance" bridging H is often described as a terminal H.

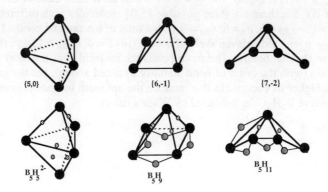

Fig. 3.19 B_5 topologies and the assignment of bonds. Small circles on edges or faces of the polyhedra indicate respectively two-centre and three-centre bonds. Terminal H's not shown; each three- or four- connected vertex has one terminal H, and each two-connected vertex has two terminal H's.

References

(1) *Handbook of Chemistry and Physics*; CRC Press: Cleveland, Ohio.
(2) Aylward, G. H.; Findlay, T. J. V. *SI Chemical Data*; John Wiley and Sons Australasia Pty. Ltd.: Sydney, 1972.
(3) Burdett, J. K. *Moleculer Shapes: Theoretical Models of Inorganic Stereochemistry*; John Wiley and Sons: New York, 1980.
(4) Rodger, A.; Johnson, B. F. G. *Inorg. Chim. Acta* **1988**, *146*, 37.
(5) Gillespie, R. J.; Hargittai, I. *The VSEPR Model of Molecular Geometry*; Allyn and Bacon: Boston, 1991.
(6) Gillespie, R. J.; Nyholm, R. S. **1975**, *11*, 339.
(7) Glidewell, C. *Inorganica Chimica Acta* **1975**, *12*, 219.
(8) Rodger, A.; Schipper, P. E. *J. Physical Chemistry* **1987**, *91*, 189.
(9) Bartell, L. S. *J. Chem. Phys.* **1960**, *32*, 827.
(10) Stoddart, J. F. *Stereochemistry of Carbohydrates*; Wiley-Interscience: New York, 1971.
(11) Deslongchanps, P. *Stereoelectronic effects in Organic Chemistry*, Pergamon Press: Oxford, 1983.
(12) Gorenstein, D. G. *Chem. Rev.* 1987, *87*, 1047.
(13) Lemieux, R. U.; N.J., C. *J. Amer. Chem. Soc.* **1958**, *133*, 31.
(14) Box, V. G. S. *Heterocylces 1990*, *31*, 1157.
(15) Jung, M. E.; Shapiro, J. J. *J. Amer. Chem. Soc.* **1980**, *102*, 7862.
(16) Barfield, M.; Dean, A. M.; Fallick, C. J.; Spear, R. J.; Sternhell, S.; Westerman, P. W. *J. Amer. Chem. Soc.* **1975**, *97*, 1482.
(17) Rodger, A.; Moloney, M. *J. Chem. Soc. Perkin Trans. 2* **1991**, 919.
(18) Kirk, D. N.; Klyne, W. *J. Chem. Soc. Perkin Trans. 1* **1974**, 1076.
(19) Greenwood, N.N. *Chem. Soc. Reviews* **1992**, 49.
(20) Wade, K. *Electron Deficient Compounds*; Nelson: London, 1971.
(21) Longuet-Higgins, H. C. *Quart. Rev.* **1957**, *11*, 121.

(22) Longuet-Higgins, H. C. *J. Chim. Phys.* **1949**, *46*, 275.
(23) Epstein, I. R.; Lipscomb, W. N. *Inorg. Chem.* **1971**, *10*, 1921.
(24) Marynick, D. S.; Lipscomb, W. N. *J. Amer. Chem. Soc* **1972**, *94*, 8699.
(25) Lipscomb, W. N. *Adv. Inorg. Chem. Radiochem.* **1959**, *1*, 117.
(26) Rodger, A.; Colwell, S. M.; Johnson, B. F. G. *Polyhedron* **1990**, *9*, 1035.
(27) Colwell, S. M.; Handy, N. C. *J. Molec. Struct. (Theochem)* **1988**, *170*, 197.

CHAPTER 4

Main Group Elements Beyond the Second Row

Contents

Introduction		95
4.1	Halogen compounds	97
	Interhalogen compounds	97
	Ionic interhalides	98
	Interhalogen oxides	99
	Group 18 (noble gas) fluorides and oxides	100
4.2	The middle of the *p* block	100
	Group 16: S, Se, Te	100
	Group 15: P, As, Sb, Bi	101
	Group 14: Si, Ge, Sn, Pb	101
4.3	The left hand side	102
	Group 13: Al, Ga, In, Tl	102
	Group 2: Mg, Ca, Sr, Ba	103

Introduction

In the previous chapter the focus was on the geometries associated with second row atoms. When consideration is extended to the rest of the main group elements it seems that in some cases nothing has changed, while in others almost nothing is the same - it all depends on which molecule one examines. In general terms, C_N's (coordination numbers) greater than four are now possible and many of the elements readily form extended arrays. As with the previous chapters, our focus will be on the local molecular geometry rather than on the extended geometries and we shall see that, for main group elements, the local geometry can usually be explained with any of the common models.

At the simplest level, we must recognise that beyond the second row of the periodic table the elements become bigger and have more valence orbitals available; in particular, the existence of unoccupied valence *d* orbitals that are energetically accessible and able to take part in bonding provides substantial flexibility to the hybridisation scheme, and hence to bond angles. As a result, the balance between electronic and steric factors shifts in favour of the steric factors as one moves beyond the second row. The consequence for predicting and systematising geometries is that both the VSEPR and the AAIM methodologies (§1.3.2) work well in most instances.

Electronic factors are still important, however. In particular, it is still electronic factors that determine the number of ligands about the central atom (up to a maximum determined by the *closo* (§2.3.1) arrangement for the given M-L bond length, usually six). If the hybridisation scheme is apparent, then a C_N follows immediately. Under such circumstances, the consequences of greater repulsion between lone pairs (VSEPR) and of L-L attraction (AAIM) are synonomous and the two methods give equivalent predictions. In general, however, predicting C_N is more complicated than for seond row atoms. There are two particular sources of difficulty.

(i) Although the s/p energy gap decreases down the periodic table and this suggests that hybridisation of s and p orbitals should get easier, the orbitals themselves get more diffuse and so give rise to weaker bonds. In extreme cases, the formation of new bonds may no longer be sufficient to pay for the involvement of s electrons in the bonding. The s electrons are then said to form *stereochemically inactive* lonepairs, which behave as core electrons even though they are in the valence shell. It is for this reason that Sn is more often found to be two-coordinate, rather than four-coordinate as tends to happen with smaller members of Group 14.

(ii) The extent to which d orbitals take part in bonding varies. This is especially true for Groups 13-18 and third row Groups 1-2, where the unoccupied d orbitals have the *same* principle quantum number as the occupied s and p orbitals and so their use in bonding may require too much energy.

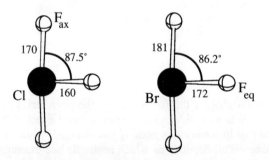

Fig. 4.1 MF_3, M = Cl, Br. Bond lengths shown in pm.

Electronic factors may also lead to orientationally-dependent bond strengths. In such cases, VSEPR is typically closer to the truth than the AAIM. An example relevant to this chapter is provided by MF_3, M = Cl, Br, which form T-shaped molecules with different axial (ax) and equatorial (eq) bond lengths (Fig. 4.1). The electronic aspects of the bonding in such molecules is considered in Chapter 5. Consideration only of the F-F interactions (AAIM) would lead to the prediction of a pyramidal {6,-3} structure. However, the hybridisation scheme required for such angles (sp^3d^2) is energetically unfavourable. In contrast, VSEPR correctly predicts the {5,-2} shape for these molecules (given that the repulsion of two lone pairs is minimised when they are equatorial rather than axial) by assuming all ten valence electrons are involved in bonding. An sp^3d hybridisation scheme provides just the requisite number of orbitals. It is interesting to note that both models contain an aspect of the truth as F-F distances are essentially close-packed [†] in accord with the AAIM (228pm for ClF_3, 241pm for BrF_3, compared with 225pm for CF_4 [1]).

The AAIM is to be preferred over VSEPR where C_N is known but the hybridisation scheme is not, or where large or small ligands (relative to bond lengths) are involved. In the rest of this chapter we will be able to proceed by ignoring electronic effects and using a balance of VSEPR and AAIM. Our aim is to give a brief overview of what geometries one might find, and to enable the observed geometries to be rationalized coherently enough so that the results can be remembered. No attempt is made to be comprehensive as the most casual glance at *e.g.* reference [2] will show. We shall use the terminology for molecular geometries developed in §2.3.1. §4.1 focuses on halogen compounds, the middle of the *p* block forms the subject matter of §4.2, and the chapter concludes with the left-hand side of the periodic table.

4.1 Halogen Compounds

With the exception of He, Ne and, for practical purposes, Ar, all elements form molecules with the halogens and these serve to illustrate the range of molecular geometries a chemist might expect to encounter. The following halides form the subject matter of this section: covalent inter-halide compounds, ionic interhalogen compounds, halogen oxides, and noble gas compounds. To clarify the parallels with the rest of this book we shall continue to use the ML_n notation, instead of the more normal practice of denoting halides by X and Y.

Interhalogen Compounds
The naturally occuring molecular forms of the halogens are F_2, Cl_2, Br_2, and I_2. The heavier elements bond in much the same manner as fluorine (Figs. 1.18-19), though the F-F bond is disproportionately weak and long. This is usually attributed to repulsion between the non-bonding electrons on the two F atoms resulting from the large electronegativity of F holding the electrons close to the nucleus and hence close to the other F atom, though it might equally be thought of as being due to the electronegative elements withdrawing electron density towards the nuclei and away from the centre of the bonds.

Differences between F and the other halogens become more apparent when the range of inter-halogen compounds is considered, since compounds of stoichiometry ML_3, ML_5 and even ML_7 exist for M = Cl, Br and I, whereas F always has a $C_N = 1$. The number of electrons in the valence shells of Cl, Br, and I limit the maximum number of covalent bonds to seven but, as it is seldom possible to fit seven atoms about a given halogen and still maintain viable bond lengths, IF_7 is the only such compound currently known. The relative stabilities of different C_N's result from a balance between stronger individual M-L bonds in

---

[†] Unconstrained bond angles significantly less than 90° are never found (§1.3.2, Table 1.1), so actual cose-packing of the F's becomes impossible with large M-L distances.

compounds with smaller C_N's, and more bonding contributions to the total with large C_N's. ClF_3 and ClF_5 are of comparable stability,[2] the decrease in bond strength being compensated by the increase in number of bonds when C_N is increased from three to five. For the less electronegative Br and I the decrease in bond strength with increasing C_N is less noticable, and so the balance shifts in favour of having as many bonds as space allows. These factors result in the following order of stability:[2]

$$ClF_3 \geq ClF_5 > BrF_5 > IF_7 > ClF > BrF_3 > IF_3 > BrF > IF_3 > IF$$

The {5,-2} geometries of MF_3, M = Cl, Br, (Fig. 4.1) were discussed above. The geometries of MF_5 are square pyramidal {6,-1} (Fig. 4.2) with the F_{ax}-M-F_{eq} bond angle decreasing from 90° to 85° to 82° along the series M = Cl, Br, I; this distortion retains fluorine - fluorine close contact as the M-F bond lengths also increase: M-F_{ax} = 162pm, 169pm, 184pm and M-F_{eq} = 172pm, 177pm, 187pm (gas phase data [2]). IF_7 is probably close packed {7,0} geometry, though some degree of distortion may exist.[2]

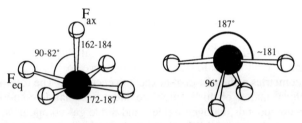

Fig. 4.2 MF_5, M = Cl, Br, I, and BrF_4^+. Bond lengths in pm, assigned as in text.[2]

It is interesting to note that HM, M = F, Cl, Br, I is the only type of halogen hydride: there are no hydrogenated mixed-halides. This is because the H-M bond is weaker than the F-M bond, and is not sufficiently strong to compensate for the energy required to hybridise the orbitals of the central halogen (§1.3.2).

Ionic Interhalides

The covalent interhalides all have odd coordination numbers and hence an even number of electrons. Attempts to create even coordination number interhalides leads to molecular ions. The geometries of ionic interhalides are only known from crystal structures and, especially for the cationic ones, they are dependent on the counterions. ML_2^-, with M = Cl, Br and L = F, Cl, or M = I and L = Cl, favour linear structures; the remaining anionic trihalides are slightly distorted from linear. In contrast, the structures of crystals containing cationic ML_2^- have two short bonds - with bond angles between 90° and 100° depending upon the counter ion - but also have two longer bonds, which makes it difficult to identify a molecular or ionic species unambiguously.

ML_4^- species all adopt a fairly close packed *trans*-{6,-2} geometry as expected from the AAIM (Fig. 1.24), or VSEPR (Fig. 1.22), whereas XY_4^+ adopts the "saw-horse" geometry illustrated in Fig. 4.2; the latter arise from electronic factors analogous to those that give rise to the T-shaped structure of BrF_3 (see above and Chapter 5) and are in accord with VSEPR.

MF_6^+, M = Cl, Br, and I are all {6,0} with twelve bonding valence electrons. BrF_6^-, however, has {9,-3} D_{3d} symmetry, which has better fluorine - fluorine interactions than the VSEPR prediction of {7,-1}. These structures were depicted in Fig. 2.15.

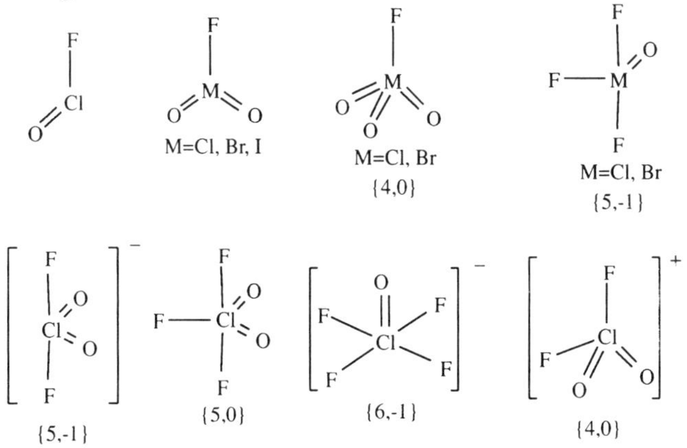

Fig. 4.3 Some oxyhalides. Data from reference [3].

Interhalogen Oxides

When an oxygen is substituted for a halogen the resulting bonds are still strong, but more of the electron density resides in the bonds than was the case with the interhalogens. This tends to minimise any electronic influences on structure and results in geometries that can be described equally well with electron repulsion (VSEPR) or L-L interactions (AAIM). MO_3^-, M=Cl, Br and I, adopt trigonal pyramidal geometries; MO_4^-, M=Cl, Br and I, are tetrahedral; and IO_6^{5-} is octahedral.

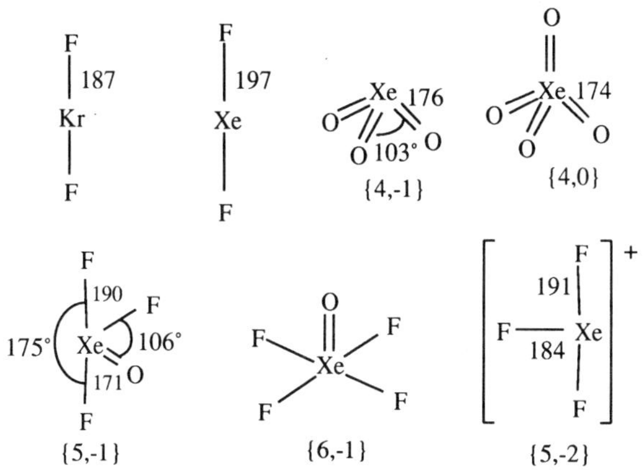

Fig. 4.4 Some noble gas, halides, oxides and oxyhalides. Data from references [1,5]. Bond lengths in pm.

A variety of geometries exist for ML_nO_m and $[ML_nO_m]^-$, M and L halogens. Some of them are illustrated in Fig. 4.3. Whenever all the ligating atom positions are not identical, the oxygens adopt the ones that maximise its π-bonding (see Chapter 5).

Group 18 (Noble Gas) Fluorides and Oxides

From the third row down, Group 18 elements have empty valence d orbitals and so have the potential to form chemical bonds. Only the chemistry of Xe is extensive and a wide range of its oxyhalides have been made; this chemistry is, however, not trivial to perform. A number of the molecules and ions that have been studied are depicted in Fig. 4.4. Once again, these geometries can be explained well in terms of both electronic repulsion (VSEPR) and L-L interactions (AAIM).

4.2 The Middle of the p Block

Group 16: S, Se, Te

The behaviour of sulfur, selenium and tellurium has some features in common with that of oxygen and some striking differences. In organic chemistry S forms a range of functional groups analogous to those of O, but S also shows the widest range of allotropes of all the elements in the periodic table and the characteristics of S-S bonds are extremely variable: S-S bond lengths ranging from 180pm to 260pm have been observed, and S-S-S bond angles vary from 90° to 180° depending on the local environment. A wide range of ML_2 (M = O, S, Se and Te) are known. Data for some of these is given in Table 4.1. The decrease in bond angles as the group is descended reflects the fact that increasing bond lengths require smaller angles for close packing of the ligands. Concomitantly there is less involvement of the valence s orbitals in bonding.

Table 4.1 Bond lengths and angles for some ML_2.

M	O	S	Se	Te
MH_2	96pm 105°	133pm 92°	146pm 91°	90°
MF_2	141pm 104°	159pm 98°		
MCl_2	170pm 111°	200pm 103°		
MBr_2	111°			251pm 98°
$M(CH_3)_2$	142pm 112°	180pm 99°	198pm 98°	

A major contrast with the behaviour of O is seen in the fluorides these elements form: S, Se and Te all form MF_6 molecules with octahedral geometries. MF_4, M = S, and Se are also known and probably adopt a {5,-1} geometry, but the experimental data is unclear; MF_5^-, M = S, Te, and SOF_4 are all {6,-1}; and SF_3^+ is pyramidal (Fig. 4.5).[2] Sulfur makes molecules with most elements of the periodic table. Se and Te bond to a wide range of elements, but these tend to form covalent solids and so it is difficult to identify molecular units. One of the more unusual molecular geometries of these compounds is the square ring such as in SNSN, Se_4^{2+} and Te_4^{2+} (Fig. 4.5).

Group 15: P, As, Sb, Bi

Nitrogen and phosphorous are ideally suited to the traditional examination question: "Compare and contrast the chemistries of ...". In addition to the increased orbital availability from the third row down, differences between P and N occur because of differences in the various bond strengths. For example, the nitrogen single bond is weaker than the P-P single bond[6] (163 kJ mol^{-1} vs. 172 kJ mol^{-1}); however, the N≡N triple bond is approximately six times stronger than the N-N single bond, while P≡P, were it to exist, would be little more stable than the P-P single bond. Thus N favours multiple bonds and P favours single bonds.[6] This trend is not confined to Group 15, as second row atoms in general tend to participate in effective π-bonding whereas for larger atoms, which have more diffuse orbitals and longer bond lengths, this is seldom attractive energetically. These electronic factors are important in determining C_N; however once that has been determined, the geometry of Group 15 molecules can be described quite adequately in terms of their non-bonded radii or VSEPR.

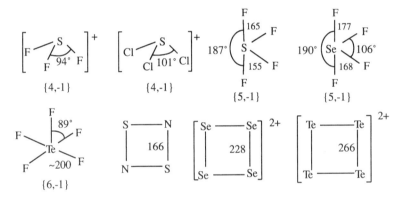

Fig. 4.5 Some Group 16 molecules. Data from references [1,2]. Bond lengths in pm.

As with Group 16 elements, various C_N's are possible for Group 15 from P down. MF_5 exists for M = P to Bi. PCl_5 and PBr_5 are {5,0} in the gas phase, though the former disproportionates into PCl_4^+ and PCl_6^- in the solid. The mixed fluorochlorides, PCl_4F and PCl_3F_2 are also trigonal bipyramidal with the smaller F in the axial positions. All the pyramidal Group 15 trihalides and trihydrides have also been observed (Table 4.2), although BiH_3 is not stable above -45°C, and BiI_3 might be better described as having C_N = 6 when in crystalline form. In general, as with Group 16, isolated molecular structures become increasingly rare down the group. Even phosphorous forms many extended structures and with a wide range of bond angles (see, e.g., reference [2]).

Group 14: Si, Ge, Sn, Pb

Silicon is the transition point between carbon, with its very well-defined covalent chemistry, and the metals below it. Its own chemistry resembles that of B (§3.3) perhaps more than that of C (§3.2), since the availability of *d* valence orbitals to take part in bonding render it comparatively electron deficient. Much of its chemistry is that of extended arrays and therefore outside the scope of this book. Si bonds with C to form molecules, e.g. benzene-like C_5SiH_6, but not with Ge, Sn or Pb. Si usually has C_N's ranging from two (where the $3s$ electrons are largely uninvolved in bonding, and bond angles are between tetrahedral and linear) to four (tetrahedral), but octahedral six coordination is also known, e.g. $[Si(acac)_3]^+$.

Table 4.2 Bond lengths and angles for some Group 14 ML_3.

M	N	P	As	Sb	Bi
MH_3	102pm 107°	142pm 94°	152pm 92°	171pm 91°	
MF_3	137pm 102°	154pm 100°	171pm 102°	203pm -	
MCl_3	176pm 107°	204pm 100°	216pm 98°	233pm 100°	248pm 100°
MBr_3		218pm 102°	233pm 101°	251pm 97°	251pm 98°
MI_3		243pm 102°	255pm 102°	267pm 99	
$M(CH_3)_3$	147pm 108°	184pm 99°	198pm 96°		

Little is known about the chemistry of germanium, but a range of halides and hydrides does exist including Ge_nH_{2n+2}, $n = 1$-5, GeL_2 and GeL_4, L = F, Cl, Br. Further down the group there is a stronger tendency towards non-bonding valence electrons, which leads to two-coordinate systems. As a result, Sn^{II} and Pb^{II} are more common than Sn^{IV} and Pb^{IV}. The range of monomeric systems is not wide though there are increasing numbers of organo tin and lead compounds being synthesised.[7] It is for the lower members of this group that the disadvantages of including s orbitals in bonding and the advantages of forming more bonds are most finely balanced, with the result that environmental factors determine which influence dominates. Accordingly, the s electrons in these two elements may be either stereochemically active, as occurs in $Sn(\eta^5-C_5H_5)_2$ (bond angle 125°), or steriochemically inactive as in $SnCl_2$ (bond angle 95°). In these examples, a satisfactory prediction of the geometry can be obtained from considering L-L interactions (i.e. using AAIM methods) without recourse to the detailed behaviour of the s valence electrons.

4.3 The Left Hand Side

Group 13: Al, Ga, In, Tl

Aluminium has some similarities to boron in its chemistry. For example trigonal planar $AlCl_3$ forms a bridged Al_2Cl_6 molecule with the B_2H_6 geometry [2] (Fig. 3.13), however, it should be noted that neither B_2Cl_6 nor Al_2H_6 are found. On the other hand, Al shows metallic behaviour and has a well defined hydrated form in water solution: $[Al(H_2O)_6]^{3+}$. Various reactive Al-based molecules exist, and have found extensive synthetic use in organic chemistry. Examples include $LiAlH_4$ and $NaAlEt_2H_2$, where the Al anion is tetrahedral, and Bu^i_2AlH which, due to the large ligands (*iso*-butyl groups), adopts a trigonal planar geometry.

Throughout the group $C_N = 1$ is common, with the valence s electrons playing little part in the bonding. Gaseous diatomic halides exist for Group 13 elements below (but not including) B. In such cases, stability increases with the size of both the cation and the anion, so that AlL (where L is a halide), GaF, and InF have been observed in the gas phase, but are unstable. Apart from these, little molecular data is available.[†]

Group 2: Mg, Ca, Sr, Ba

Isolating molecular geometries for Group 2 elements is not easy since they generally form ionic bonds which are stable either in lattices, or in solution as dissociated ions. However, in solution they do form definite complexes with water, and calcium and magnesium are widespread in biological systems, usually complexed to other molecular or ionic units[8], see *e.g.* Fig. 7.11.

As far as molecular geometry discussions go, the alkali earth halides provide a series of compounds which illustrate clearly the failure of VSEPR theory for "electron deficient" systems. BeF_2 and MgF_2 are linear but CaF_2, SrF_2 and BaF_2 are bent, illustrating the effects of L-L attraction (AAIM). The full table of Group 2 halide bond angles is given in Table 4.3.

Table 4.3 Bond angles for Group 2 halides, ML_2. Data from reference [9].

M	L	F	Cl	Br	I
Be		180°	180°	180°	180°
Mg		155°-180°	180°	180°	180°
Ca		133°-155°	180°	173°-180°	180°
Sr		108°-135°	120°-143°	133°-180°	161°-180°
Ba		100°-115°	100°-127°	95°-135°	102°-185°

Under some circumstances Group 2 elements can also achieve an octet of valence electrons and a tetrahedral geometry. Examples include: $Be(OEt_2)_2Cl_2$, $[NEt_4]_2[MgCl_4]$, and $Mg(OEt_2)BrEt$, where Et is an ethyl group. $BeCl_2$ also manages to increase the valence electrons about the beryllium, in this case by dimerising to a planar structure with two Cl's bridging between the Be's.

References

(1) Burdett, J. K. *Moleculer Shapes: Theoretical Models of Inorganic Stereochemistry*; John Wiley and Sons: New York, 1980.
(2) Greenwood, N. N.; Earnshaw, A. *Chemistry of the Elements;* 1st ed.; Pergamon Press: Oxford, 1984.
(3) Christie, K.O.; Schack, C.J. *Adv. Inorg. Chem. Radfiochem.* **1976**, *18*, 319; Downs, A.J.; Adams, C.J. in *Comprehensive Inorganic Chemistry*, 2, 1386, Bailar, J.C.; Eméleus, H.J.; Nyholm, R.S.; Trotman-Dickerson, A.F. (eds), Pergamon Press: Oxford, 1973.

[†] Note that the molecular formula can often be misleading for these molecules, *e.g.* TlI_3 is in fact Tl^I not Tl^{III}.

(4) Aylward, G. H.; Findlay, T. J. V. *SI Chemical Data*; John Wiley and Sons Australasia Pty. Ltd.: Sydney, 1972.
(5) Huston, J.L. *J. Inorg. Chem.*, **1982**, *21*, 685.
(6) Purcell; K.F.; Kotz, J.C. *Inorganic Chemistry,* Holt-Saunders International, 1977.
(7) Pereyre, M.; Quintard, J. -P.; Rahm, A. *Tin in Organic Synthesis,* Butterworths: London, 1987.
Block, E. (ed.) *Heteroatom Chemistry,* VCH Publishers: New York, 1990.
(8) Williams, R. J. P. *The Biological Chemistry of the Elements,* Clarendon Press: Oxford, 1991.
(9) Gillespie, R. J.; Hargittai, I. *The VSEPR Model of Molecular Geometry*; Allyn and Bacon: Boston, 1991.

CHAPTER 5

Complexes of Transition Metals and f-Block Elements

Contents

Introduction			106
5.1	Transition metal complexes		107
	5.1.1	A survey of transition metal complexes by coordination number	107
		$C_N = 2$	108
		$C_N = 3$	108
		$C_N = 4$	108
		$C_N = 5$	109
		$C_N = 6$	109
		$C_N = 7$	110
		$C_N = 8$	110
		$C_N = 9$	110
	5.1.2	Determining transition metal complex geometries: an overview	111
	5.1.3	Crystal field theory	112
		Assignment of electrons	113
		Inadequacies of crystal field theory	116
	5.1.4	Ligand field theory	116
		Transition metal complexes with no M-L π interactions.	117
		Transition metal complexes with σ and π M-L	117
		The magnitude of Δ and the eighteen-electron rule for $\{6,0\}$ ML_6	118
		The eighteen-electron rule and non-octahedral systems	119
	5.1.5	Steric versus electronic effects on transition metal complex geometry	120
		Which coordination number?	120
		Which template?	121
		Steric interactions in *tris*-chelate complexes	122
		Racemization reactions of *tris*-chelate complexes	125
		The *trans*-effect and the *trans*-influence	127
		The Jahn-Teller effect	130
		The geometry of metal hexaquo systems	131
5.2	Lanthanides and actinides		132

Introduction

The subject matter of this chapter is the geometries adopted by metal complexes. A metal complex may be defined as a molecule with n ligands, L, bonded directly to a metal atom or ion, M, *via* the donation of two electrons into a M-L σ bond. The ligands may be atoms, molecules, or ions and usually donate two electrons to the M-L bond, so that the metal therefore nominally has its own plus $2n$ electrons in its valence shell. There may additionally be π or even δ[†] bonding interactions. Ligands commonly bond to M through an N, O, S or a halogen atom, but other possibilities do exist.

Simple approaches to molecular geometry, such as VSEPR theory (§1.3.2), founder when they encounter transition metals. Most text books therefore study transition metal complexes using ligand field theory (LFT). LFT results from incorporating covalency into the electrostatic crystal field theory (CFT) using molecular orbital (MO) (§1.3.1) theory (or to be more precise, MO ideas). The slide from CFT to LFT is usually blurred and this can leave one with a confused view of transition metal geometry determination, being unsure of what assumptions belong where.

As with main group systems, the geometry arises as a compromise between electronic and steric factors, although the position of balance shows a much richer variety than tends to be the case for the main group elements. The focus of this chapter will be on the additional geometric considerations raised by the availability for bonding of d (transition metal) and f (lanthanide and actinide) valence orbitals and electrons. Our approach will be consistent with that used for main group systems. The subject matter is vast and we shall make no attempt to be comprehensive. Our aim, as in previous chapters, is to provide a framework that will allow the reader to understand and utilise other books and articles in chemical journals.

Most space in this chapter is given to transition metal complexes because they form so many more compounds than do the f block elements; also, rationalising their geometry provides much more of a challenge for the chemist. Before discussing the particular electronic and steric features of transition metal complexes, a brief survey of the range of C_N's (coordination numbers), and geometries found for transition metal complexes (in the sense of molecular units, rather than parts of extended lattices) is given. Although six-coordination with an octahedral geometry is by far the most common geometry, there is more variety than a casual glance at an inorganic chemistry textbook might indicate. This chapter runs the risk of shifting the emphasis too far in the direction of the more unusual cases, so should be read in parallel with a standard text book.[1-5] CFT (§5.1.3) and LFT (§5.1.4) are used to describe the electronic features of transition metal complexes in as simple a way as possible without being completely misleading. A discussion of steric and electronic factors follows in §5.1.5 concluding with some stereoelectronic effects of square planar complexes and the Jahn-Teller effect. The final section of the chapter contains a brief discussion of molecules of the f-block elements (§5.2).

[†] A δ bond is formed *e.g.* when two d orbitals overlap face to face making a bond with two nodal planes along the bond axis.

We shall again make use of the structural notation developed in §2.3.1. Thus, *e.g,* an octahedral six coordinate system is *closo* and denoted {6,0}, whereas a square planar geometry is based on that octahedral template but has two vacancies and so is denoted *trans*-{6,-2}, or simply {6,-2}. It is not always clear how to label a structure. For example, the saw-horse geometry adopted by ClF_2O_2 (Fig. 4.3), might be either distorted {5,-1} or distorted *cis*-{6,-2}, depending on the O-Cl-O bond angle. However, in mosts cases the notation is unambiguous and simple.

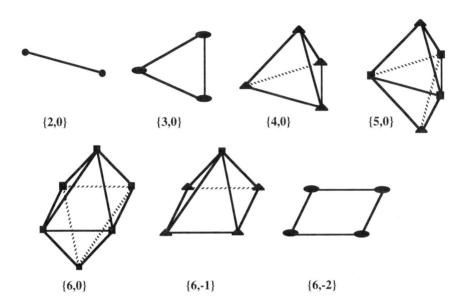

Fig. 5.1 Common ligand-polyhedron geometric templates used to describe ML_n geometries, $n = 1$-6. Metal atoms are placed equidistant from all ligands if possible. Some common derivative *arachno,* and *nido* geometries are also illustrated. Vertices denote ligands and are labelled with shapes indicating the number of nearest neighbour ligands.

5.1 Transition Metal Complexes

5.1.1 A Survey of Transition Metal Complexes by Coordination Number

Transition metal complexes adopt C_N's ranging from two to nine. Examples of each type are given below, and are discussed in somewhat more detail in standard inorganic textbooks, *e.g.* references [1-5]. The geometric templates used to describe metal complex geometry are illustrated in Figs. 5.1-3 and were discussed in some detail in §2.3. This section is intended as a survey of the types of geometries that can arise. More detailed discussion of why a given geometry is observed follows later in the chapter.

$C_N = 2$

Most such complexes involve d^{10} metals such as Cu(I), Ag(I) and Au(I). All known complexes are linear with $D_{\infty h}$ symmetry. Examples include $[CuCl_2]^-$; $[Ag(NH_3)_2]^+$; $[Ag(CN)_2]^-$ and $[Au(CN)_2]^-$. Two coordination is also possible for M = Mn, Co, Ni, Zn, Cd, Hg, e.g. Ni(Si(CH$_3$)$_3$)$_2$, and presumably other two-coordinate species are possible if the ligand is sufficiently bulky.

$C_N = 3$

This coordination is rare. Systems such as $[HgI_3]^-$ and $[HgBr_3]^-$ exist, (though Hg^{2+} is not strictly a transition metal); the latter is an example of a non-planar three coordinate complex, having the Hg 32pm above the plane of the Br atoms. More generally, three-coordination arises as a result of specific steric restrictions. For example, bulky ligands give rise to the trigonal planar {3,0} complexes Fe{N(Si(CH$_3$)$_3$)$_2$}$_3$, Cu{SC(NH$_2$)$_2$}$_3$, and [Cu(SPPh$_3$)$_3$]ClO$_4$, where Ph denotes phenyl. Steric restrictions may also arise from bonding considerations in multidentate ligands, so that AgCH(CN)$_3$ forms a trigonal pyramidal {4,-1} rather than {3,0} due to coordination being through the three N atoms in CH(CN)$_3$. Finally, three-coordination may also occur when there is a deficiency in the number of ligands available for complexation, and it is this that leads to the formation of FeCl$_3$ as a planar complex in the gas phase.

$C_N = 4$

Molecules with the T_d {4,0} or D_{4h} trans-{6,-2} geometries are common in metal complexes. The full range of geometries intermediate between these two are also observed. {4,0} is particularly common with d^5 and d^{10} systems such as $[MnO_4]^-$, Ni(CO)$_4$ and [Cu(pyridine)$_4$]$^+$, since the bonding can be accomplished using just s and p orbitals and so leaves a spherical d shell. Cu(II) molecules also form T_d complexes, though a Jahn-Teller distortion (§5.1.5) usually results in some distortion towards a square planar geometry. For electronic reasons (§5.1.5) d^8 complexes favour square planar geometries. {6,-2} Pd(II) and Pt(II) systems are often found. d^8 Ni(II) complexes adopt both four coordinate geometries and intermediate ones between them (§5.1.5). Metals with seven or fewer d electrons have smaller crystal field stabilisation energies (§5.1.3) and tend to favour higher C_N's, especially six. However, low positive charges on the metal coupled with anionic ligands that are low in the spectrochemical series (§5.1.3) may lead to tetrahedral or distorted tetrahedral geometries; thus $[CoCl_4]^{2-}$, $[FeCl_4]^{2-}$ and $[MnI_4]^{2-}$ are found. As with three- and two-coordination, bulky ligands may force a {4,0} geometry on a metal that would normally adopt higher C_N's, e.g. [Fe(OPPh$_3$)$_4$]$^{2+}$ and Cr(OBu^t)$_4$ both have T_d symmetry.† Tetradentate analogues of the tridentate CH(CN)$_3$ example for three-coordination are far harder to engineer, as they now require the ligand to encapsulate the metal; usually this level of sophistication in transition metal complexes is only found in biological enzymes.

---

† Ph = phenyl, and Bu^t = tertiary butyl.

$C_N = 5$

Five was originally (nineteenth and early twentieth centuries) thought to be a very unusual C_N, but has since been shown to be quite common, especially for d^7, d^8, and d^9 metals. Bulky rigid ligands may preclude the approach of a sixth ligand resulting in $C_N = 5$; however there are more "natural" examples such as $[CuCl_5]^{3-}$; $[M(CN)_5]^{3-}$, M=Co, Ni; $[MnF_5]^{3-}$; and $CoCl_3(PEt_3)_2$. Geometries for these complexes derive from the trigonal bipyramidal {5,0} and square-based pyramidal {6,-1} templates, though usually the observed geometries fall somewhere between these two extremes. Even so, it is more common for geometries to be close to {5,0}, since the vacancy in {6,-1} forms a ready site for the addition of another ligand unless the metal already has eighteen electrons (§5.1.4). Mixed ligand systems such as $VO(acetylacetone)_2$ form {6,-1} geometries, in this case, with the O at the apex.

Five-coordinate complexes provide some good examples of fluxional behaviour. A fluxional system is one where rearrangements (*cf.* §2.3) occur on a timescale that is fast compared with the timescale of the experiment. Both {5,0} and {6,-1} have two different ligand environments and this should be shown by experimental probes such as NMR. However, in many cases the system is very labile, and ligands rapidly exchange between the different types of sites. A good example is $Fe(CO)_5$, which is trigonal bipyramidal but shows five equivalent ligands on the NMR timescale (10-100ms).

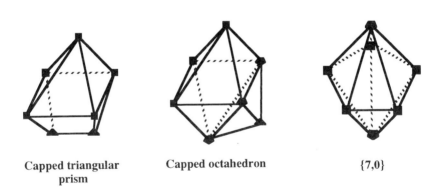

Capped triangular prism   **Capped octahedron**   {7,0}

Fig. 5.2 Common ligand-polyhedron geometric templates used to describe ML_7 geometries oriented to show the interconversion between them. Notation as for Fig. 5.1.

$C_N = 6$

The discussions of electronic structure and steric factors given below are biased towards six-coordinate systems and this reflects the dominance of this geometry in transition metal complexes for both electronic and steric reasons. By far the most common six-coordinate geometry is octahedral, and we have made some attempt to convey this by choosing a wide range of octahedral complexes to illustrate various points in the following sections. Other six coordination geometries do occur, the most common being {9,-3}, the trigonal prism, which is found in *tris*-chelate complexes (see §2.3.3 and §5.1.5).

$C_N = 7$

Seven coordination is rare as it tends to be an inefficient way of packing ligands around the metal: either there is not enough room to accommodate the seventh ligand, or a small rearrangement of the ligands would readily provide the room to accommodate still more ligands. Such examples as do exist tend to involve small ligands, and those second and third row transition metals that have the larger valence-orbital radii but few d electrons. Examples include $[MF_7]^{3-}$, M = Zr, Hf; and $[MF_7]^{2-}$, M = Nb, Ta. The three possible high-symmetry templates associated with this C_N - the pentagonal bipyramid, {7,0}, the capped octahedron, {8,-1}, and the capped trigonal prism, {9,-2} - require only small distortions to change from one to the other (Fig. 5.2). This is especially true for the latter two. $[MF_7]^{3-}$ are usually {7,0}, whereas $[MF_7]^{2-}$ are usually {8,-1}, presumably reflecting the lower L-L repulsion in the latter case where the charges on the ligands will be lower.

$C_N = 8$

Eight coordination is even rarer than seven in isolated molecular or ionic units; it is more common in ionic crystals such as CaF_2 where the electrostatic rather than bonding considerations determine the geometry. Square antiprismatic {10,-2}, examples are $[TaF_8]^{3-}$, $[W(CN)_8]^{4-}$, and $Zn(acetylacetone)_4$. The geometry adopted by $[ZnF_8]^{4-}$, $[Mo(CN)_8]^{4-}$, and $[Co(NO_3)_4]^{2-}$ is the triangulated dodecahedron, {8,0}. The cube has yet to be observed as it leaves too much free space around the transition metal for the geometry to be particularly stable.

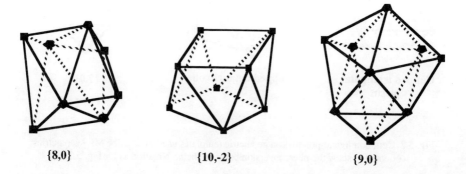

{8,0} {10,-2} {9,0}

Fig. 5.3 Common ligand-polyhedron geometric templates used to describe ML_n geometries, $n = 8,9$. Notation as for Fig. 5.1.

$C_N = 9$

$[ReH_9]^{2-}$ adopts a {9,0} geometry.

5.1.2 Determining Transition Metal Complex Geometries: An Overview

We have just seen that metal complexes exist with C_N's ranging from two to nine. So the first question is: for a given metal and ligand what coordination might be expected? Then we must ask how the ligands are arranged about the metal. We have already mentioned that transition metal geometries result from a balance between electronic and steric influences. So if we are to understand their geometries, we need to understand the underlying electronic and steric influences first. A great deal of theory and chemical intuition has been developed in this field, and it is important to understand the concepts involved if the literature on transition metal chemistry is not to be simply a magical black box. We shall begin by considering the electronic structure of an isolated transition metal, and then examine the way this can be modified by its environment.

Before proceeding with this outline, it is worth noting two experimental observations about the electronic structure in transition metal complexes. The first is that although in an isolated atom the valence s orbitals lie below the valence d orbitals (hence the periodic table, §1.2) this order is reversed in the presence of any ligands. Thus the s orbitals will be unoccupied for oxidation states greater than +2 in transition metal complexes, and may even be unoccupied for oxidation states of +1: Cu^+ shows a spectroscopy that is consistent with d^{10} rather than s^1d^9. The second is that the spectroscopy of the d energy levels shows only small perturbations away from what would be expected for the isolated transition metal ion. The transition energies and intensities, and the selection rules are all compatible with orbitals that are predominantly transition metal d character. This is not to say that there is no interaction with the ligands, but that it is small compared with the bonding interactions involving s and p orbitals. A consequence of this is that any transfer of electron density between the metal d orbitals and the ligands is also relatively small.

In isolation, a transition metal atom, M, has five energetically degenerate d orbitals. The individual orbitals are distributed in space, but the net d electron density distribution is spherical since the electrons on average occupy the five degenerate orbitals equally. As soon as M is placed in the non-spherical environment provided by ligands, the d orbitals become non-degenerate[†] and differently shaped, and oriented d orbitals are differentially occupied. For example, consider the effect on the M d orbitals of the approach of an L with two electrons directed towards the metal (we assume the ligand electronic structure has been organised so that it is pre-aligned to make a σ bond with the metal). Using the fragment *mo* approach (§1.3.1) we can see how the valence orbitals of metal and ligand are changed to form those of the new ML system. Any d orbital along the line of approach will make a bonding and an antibonding interaction with the occupied ligand orbital from which the electrons or the M–L bond will come. If the d orbital is already occupied, then a net destabilisation will occur. Thus the orientation of the ligands about the metal in a transition metal complex is such as to reduce repulsion between the incoming ligand electrons and existing d

---

[†] Transition metals are too small and do not have enough valence orbitals to have twelve icosahedrally coordinated ligands so cannot have **I** or **I**$_h$ symmetry so the d-orbitals cannot be degenerate.

electrons (though it should be noted that most of the bond strength actually comes from interaction with metal valence s and p orbitals). The "shapes" of the five d orbitals were illustrated in Fig. 1.15.

How many such ligands may be accommodated about M? The simplest prediction of C_N comes from the eighteen-electron rule, which states that stable transition metal complexes will have eighteen valence electrons. Its justification lies in the fact that first and second row transition metals have nine valence orbitals. Hence, by analogy with the main group eight-electron rule (Chapter 3), eighteen valence electrons should be optimal to ensure nine occupied bonding orbitals and no occupied anti-bonding orbitals. Examples in support of this rule are easy to find, $e.g.$ $[ZnCl_4]^{2-}$, $Fe(CO)_5$, and $[Co(NH_3)_6]^{3+}$. However, it does seem to be one of those rules that is more honoured in the breach than in the observance, and complexes with metal electron counts ranging from twelve to twenty-two are common (§5.1.4).

The first reason for the eighteen-electron rule being broken is steric: there may not be sufficient room to accommodate enough ligands to provide the electrons. Thus, $Fe\{N[(CH_3)_3]_2\}$ has a fourteen-electron count and metals on the left hand side of the periodic table (i.e. those that need more electrons from the ligands) often have a low count. Electronic factors can also be important and may override the eighteen-electron rule. If by adding another ligand the M-L bond energy so gained is greater that the destabilising interaction of accommodating more than eighteen electrons about the metal, then higher electron counts will be found. We shall return to this below in the context of LFT, since the theory can be used to understand when the eighteen-electron rule may be expected to hold.

5.1.3 Crystal Field Theory (CFT)

CFT was the first successful method for treating environmental effects of ligands on the d-electrons of a transition metal. It originated in the context of ionic crystals, and the ligands were treated as point negative charges that repel d electrons and destabilise d orbitals. LFT (see below) follows from CFT when it is acknowledged that the bonding in metal complexes has a significant degree of covalent character. CFT does not consider the details of bonding interactions - all that matters is the change in the energy of the d electrons due to interacting with the anionic ligands.

A number of conventions are adopted in deriving d orbital energy level diagrams from CFT. All ligands are taken to be equivalent and to cause one unit of destabilisation (the magnitude of this unit depends on the identity of the M and L, see below). The CFT "zero" energy for ML_n is taken to be the d orbital energy that would arise if the electric field due to the ligands were spherical, i.e. $n/5$ units above the isolated ao "zero" since the total destabilisation would be shared by five d orbitals. Thus, in the CFT energy level diagrams (Fig. 5.4), it appears that some orbitals are stabilised and some destabilised; however, this is only with respect to the CFT "zero", not with respect to the isolated atoms. The lowest orbitals are those that are assumed to be unaffected by the presence of the negative charges, i.e. the ones that do not point towards the anions. The crystal field stabilisation energy (CFSE) $\Delta = 10D_q$ is defined to be the maximum energy gap between d

orbitals in the CFT energy level diagram. Thus, the well-known $\Delta_{oct} = 9/4\Delta_{tet}$ relationship is due to the facts that six ligands cause more destabilisation than four, and in {4,0} three d orbital share the destabilisation in contrast to the situation for {6,0} for which only two orbitals share the destabilisation.

In order to quantify these effects, as done for Fig. 5.4, we need to know how the one unit of destabilisation caused by each ligand is spread amongst the d orbitals with which it interacts. CFT assumes this partition is proportional to the density each orbital has in the region of the ligand (assuming fully occupied d orbitals). Thus a ligand situated on the z axis destabilises d_{z^2} by one unit. However, one situated on the x axis affects both $d_{x^2-y^2}$ and d_{z^2}. The relative destabilisation by such an x-directed ligand comes from the ratio of the square of the x^2 coefficient in the angular distributions of the orbitals; thus using coefficients given in Fig. 5.4, [destabilisation of $d_{x^2-y^2}$]/[destabilisation of d_{z^2}] = $(1/\sqrt{2})^2/(1/\sqrt{6})^2 = 3$. As the total destabilisation from one such ligand is one, it destabilises $d_{x^2-y^2}$ by 3/4 and d_{z^2} by 1/4. The answer is less obvious when the approach is not aligned with an orbital. For {3,0} (and also {7,0}, axial-{7,-1} and axial-{7,-2}) we use symmetry and note that $d_{x^2-y^2}$ and d_{xy} are degenerate by symmetry (so equivalently distorted by the ligands in the x-y plane) and when combined their electron density would form a toroid about the x-y plane (analogous to the middle of the d_{z^2} orbital). Thus the three ligands which are in the x-y plane destabilise $d_{x^2-y^2}$ and d_{xy} equally, and the ratio of their *combined* destabilisation to that of d_{z^2} is the ratio of the sizes of their toroids, namely $(1/\sqrt{2})^2/(1/\sqrt{6})^2 = 3$, again using the coefficients in Fig. 5.4; so three ligands destabilise d_{z^2} by 3/4 and each of $d_{x^2-y^2}$ and d_{xy} by 9/8. The ligands for {9,-3} are placed about 45° from the z axis so they interact only with d_{xz} and d_{yz}. (If all M-L bond lengths and L-L distances are the same the z-axis to M-L angle would be $\cos^{-1}(\sqrt{(3/7)}) = 49°$.)

Assignment of the Electrons

Once an orbital energy level diagram has been determined, the next stage is to assign electrons to orbitals to determine the ground electronic configuration and state. This is not as straight-forward as with main group systems since the energy splitting between the d orbitals, Δ, is quite small, and can sometimes be less than the energy, P, of pairing the spins of two electrons in the same orbital. Electrons are assigned in accord with the aufbau principle and Hund's rules as with main group systems until the first d electron that would be spin paired is being considered. If Δ is large the next electron follows main group behaviour and occupies one of the half-filled orbitals with spin opposite to the electron already there; however, if Δ is smaller than P then the next electron will go into a higher energy unoccupied d orbital with spin parallel to the existing electrons. Thus, for example, for square planar $\mathbf{D}_{4h}$ with four d electrons, the electrons may adopt either a low spin or a high spin configuration as illustrated below.

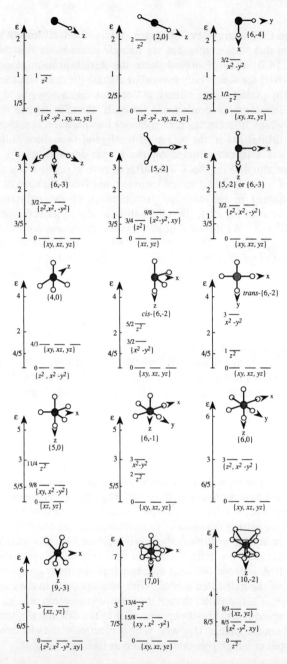

Fig. 5.4 CFT energy level diagrams for d orbitals in a number of common ML_n geometries using the isolated ao's as the energy "zero". The CFT "zero" of $n/5$ is shown. z is taken to be the unique bond axis if one exists. The five d orbitals are labelled in the figure by their cartesian labels, with z^2 standing for $d_{2z^2-x^2-y^2}$. All ligands are taken to be equivalent. The normalised cartesian forms of the angular distributions of the d orbitals are $N(x^2-y^2)/\sqrt{2}$, Nxy, Nxz, Nyz, and $N(2z^2-x^2-y^2)/\sqrt{6}$ where $N=\sqrt{(15/16\pi)}$.

Some geometries, such as {4,0} have small values of Δ for all ligands and so always adopt high spin configurations for the ground electronic state. On the other hand, octahedral complexes adopt both high and low spin configurations depending on the metal, and on the strength of the crystal field (or more accurately, depending on the net value of Δ as determined by all interactions between the metal and the ligands, as discussed below). P ranges from 19,000cm^{-1} for Fe^{2+} to 30,000cm^{-1} for Fe^{3+}.[†,6,7] Δ has a greater range of values, depending upon both M and L. Water give rise to an intermediate-strength crystal field, and for hexaquo complexes of the first row transition metals Δ ranges from 5,500cm^{-1} for Mn^{2+} to 21,000cm^{-1} for Mn^{3+}.[8,9] The ability of various common ligands to split the d orbitals has been determined and their ranking in terms of increasing Δ values is called the spectrochemical series for the ligands:

$I^- < Br^- < Cl^- < SCN^- < F^- < OH^- <$ acetate $<$ oxalate $< H_2O <$
$NCS^- <$ glycine $<$ pyridine, $NH_3 <$ ethylenediamine $< SO_3^{2-} <$
bipyridine, phenanthroline $< CN^-$

An approximate metal ion spectrochemical series is:

$(Mn^{2+} < Ni^{2+} < Co^{2+} < Fe^{2+} < V^{2+}) < (Fe^{3+} < Cr^{3+} < V^{3+} < Co^{3+}) <$
$Mn^{4+} < (Mo^{3+} < Rh^{3+} < Ru^{3+}) < Pd^{4+} < Ir^{3+} < Re^{4+} < Pt^{4+}$

where the parentheses group elements that are close together in the series.

As a general rule, second and third row transition elements are low spin. This is because the d orbitals are more diffuse for these elements than for the first row (higher principle quantum number) so that electrons in these orbitals will be further from the metal nucleus. As a result they will spend more time closer to the ligands and so interact more strongly with them. Hence Δ values will be larger. Octahedral Co^{3+} complexes are also almost always low spin, whereas most other first row transition metals 3+ ions are high spin. Octahedral first row transition metal 2+ ions are generally high spin unless coupled with ligands very high in the spectrochemical series.

It should be noted that the whole high spin / low spin argument runs the danger of being a little circular. High spin systems tend to have longer (and hence weaker) M-L bonds compared with similar low spin systems, since the ligands are repelled by the electrons in the destabilised d orbitals (e_g^*: d_{z^2} and $d_{x^2-y^2}$ for O_h systems). This, in turn, causes Δ to be smaller as the ligands do not destabilise the e_g^* electrons as much as if the bond lengths were the same as for low spins systems and so makes high spin configurations more likely.

There is an interesting consequence of this bond-lengthening process when switching from low- to high-spin states. From the preceding discussion, one can expect that electron transfer reactions between +3 and +2 oxidation states for first row d^4 - d^7 transition elements will often involve a change of spin state and so will be accompanied by a significant change of bond length. Whereas the transfer of a single electron could be a very rapid process, a change that also involves

---

† Because most data for P and Δ come from spectroscopy, values tends to be given as a wavenumber rather than as an energy, but the two are equivalent with 1cm^{-1}(molecule)$^{-1}$ ≡ 11.97Jmol^{-1}.

movement of the much heavier nuclei will be a much slower process. Thus electron transfer rates can be used to deduce geometric information.

Inadequacies of Crystal Field Theory

We have managed to derive a fair semblance of an orbital energy level diagram for the d orbitals (Fig. 5.4) of a transition metal complex by representing the ligands as point negative charges. This is a simplification, however, as there will always be some degree of mixing between the d orbitals of the metal and the σ and π *mo's* of the ligand. Although such interactions are not likely to be large - the d orbitals tend to be close to the nucleus and so do not give appreciable overlap with orbitals from the ligands - they do lead to quantitative changes from the CFT predictions of the orbital energies. The differences can be very important for determining the geometry, as well as the spectroscopy, reactivity, and magnetism, of transition metal complexes.

5.1.4 Ligand Field Theory (LFT)

LFT is an extension of CFT designed to accommodate the M-L orbital overlap effects that are explicitly ignored in CFT. It is based on the MO and fragment-MO ideas of bonding that were developed in §1.3.1. The ligands are represented, not by negative point charges as in CFT, but by occupied and unoccupied orbitals available for bonding to M. We have already begun to develop LFT when, in Chapter 2, we used symmetry to derive approximate *sao's* (symmetry adapted orbitals) for $[MnCl_6]^{4-}$ (§2.2.1). Each Cl^- was first represented by a p or sp hybrid orbital oriented along the M-L bond ready to interact with the Mn valence orbitals to make a σ bond (Fig. 2.11). We then added Cl^- orbitals perpendicular to the Mn-Cl bonds that were oriented to overlap with the d_{xy}, d_{xz}, d_{yz} orbitals to make in π bonds. The qualitative MO energy level diagram that resulted was shown in Fig. 2.12.

Ligands may be classified according to the types of bonds they make with M. All ligands are σ-donors. Each NH_3 in $[Co(NH_3)_6]^{3+}$, for example, has an occupied lone pair orbital that is oriented along the M-L bond and takes part in σ interactions. The result is Co-N bonds that are composed mainly of N orbitals but have some metal $4s$, $4p$, $3d_{x^2-y^2}$, and $3d_{z^2}$ character, with each bonding orbital occupied by two electrons. NH_3 has no other available valence electrons or orbitals. Other ligands, such as Cl and CO have aditional orbitals available that are suitably oriented to make π bonds with (in octahedral {6,0}-ML_6 molecules) metal d_{xy}, d_{xz}, and d_{yz} orbitals. In the case of Cl, the only available orbitals are occupied and interaction with M orbitals does lead to some transfer of electron density to M. Such ligands are referred to as π-donors. CO, however, has both occupied and unoccupied π orbitals. The unoccupied ones lie slightly above the M d orbitals, and extend more towards M than do the occupied ones (see shapes of diatomic orbitals in Fig. 1.18). The d orbitals with which they interact are occupied and therefore some transfer of electron density occurs from M to L into CO π* orbitals. (A side effect of this interaction is that the CO bonds are weakened.) Such ligands are referred to as π-acceptor ligands.

Our focus in the following discussion will be to see how the d-orbital part of the diagram changes according to the ligand and geometry involved. It is rather ironic that most of the bond strength in transition metal complexes arises from interactions with M orbitals other than the d orbitals, yet the d orbitals seem to have most effect on the geometry observed. This is largely due to the fact that the lowest occupied and highest unoccupied *mo's* of transition metal complexes are (more-or-less by definition) orbitals with mainly d character.

Transition Metal Complexes with no M-L π Interactions

In any transition metal complex the L σ orbitals lie below the M^{n+} valence levels. Thus the M-L bonding orbitals will have more L than M character, and conversely for the antibonding orbitals. This has a number of consequences. Firstly, the M d orbitals will be destabilised by the σ interaction with the L's and give rise to a splitting, Δ, of the d orbitals that is qualitatively similar to the CFT diagrams (Fig. 5.4). Secondly, although the bonding orbitals are predominantly L in character, they do contain some contribution from M, and so there will be a (small) net transfer of electron density from ligands to the metal; this is called a metal to ligand "dative" bond.

Transition Metal Complexes with σ and π M-L Interactions

The main difference from the previous case is that some combination of the available π L orbitals has the correct symmetry to interact with the d orbitals that are not involved with σ M-L bonds. For O_h {6,0} systems this means the three t_{2g} orbitals: d_{xy}, d_{yz} and d_{xz}. This interaction can either raise or lower the energy of the d orbitals, depending on whether they have higher or lower energy than the π L orbitals, which in turn depends on whether or not the π L orbitals are occupied as discussed above.

π-donor ligands: π-donor ligands have occupied π L orbitals lower in energy than the d orbitals as discussed above. Interaction with M orbitals results in bonding orbitals that are lower in energy than the original π orbitals and have a small amount of M character, and antibonding orbitals (of t_{2g} symmetry for {6,0}) with mainly d character that are higher in energy than in the absence of π bonding. These antibonding orbitals may or may not be occupied, depending upon the number of d electrons. The d orbital energy changes are illustrated in Fig. 5.5. The energy scale on this figure is much smaller than for the corresponding σ diagram, Fig. 5.4. The two diagrams should be combined to deduce the net value of Δ. This is illustrated for {6,0} in Fig. 5.6.

π-acceptor ligands: π-acceptor ligands have unoccupied π^* L orbitals higher in energy than the (partially) occupied d M orbitals. Interaction with M orbitals results in bonding orbitals (t_{2g} for {6,0}) that are lower in energy than the original d orbitals that have mainly d character, and antibonding L orbitals with a small amount of M d character that are higher in energy than the original π^* L orbitals. The effect of a π-acceptor ligand on the d orbitals in a number of different

geometries may be found by turning Fig. 5.5 upside-down. Fig. 5.6 shows the effect of a π-acceptor under octahedral symmetry.

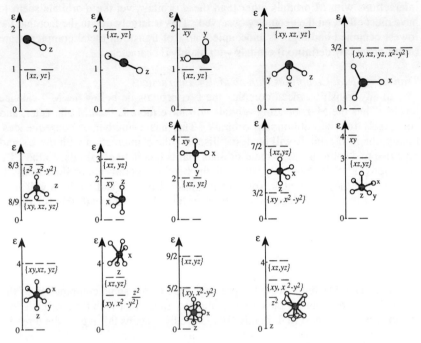

Fig. 5.5 Extra destabilisation of d orbitals in a number of common ML_n geometries with π-donor L. "Zero" is defined by the energy of the isolated $ao's$. Labelling is as for Fig. 5.4 but only destabilised orbitals are indicated explicitly. The net destabilisation of a d orbital is determined by noting that the ligand π orbitals are perpendicular to the M-L bond. We assume each ligand has two π orbitals each giving one unit of destabilisation to the d orbitals with which they overlap. But note that the π* orbitals have two lobes of opposite sign so that no net overlap between d_{z^2} and π ligands in the x-y plane occurs. The diagram for {4,0} follows from noting that $d_{x^2-y^2}$ and d_{z^2} are degenerate by symmetry, as are d_{xy}, d_{xz} and d_{yz}, and that four d_{xy} lobes overlap with π orbitals, whereas d_{z^2} has the equivalent of twelve since its xy electron density covers the x-y plane uniformly; those for {9,-3} and {10,-2} are only approximate. More accurate diagrams may be determined using the angular overlap model.[10,11] The diagram for π-acceptor ligands follows by reflecting othe energy scale about "zero", so that E increases from top to bottom.

The Magnitude of Δ and the Eighteen-Electron Rule for {6,0} ML_6

We are now in a position to understand when the eighteen-electron rule might be expected to hold. For O_h {6,0} complexes there are eleven orbitals that, if occupied, provide electrons for the M valence shell: the six σ-bonding orbitals, which are mainly ligand in character, and the five d orbitals. In the absence of any M-L π bonding the e_g d orbitals ($d_{x^2-y^2}$ and d_{z^2}) are antibonding and the t_{2g} d orbitals (d_{xy}, d_{yz}, and d_{xz},) are non-bonding (Fig. 5.4). If L is a strong σ-donor, then Δ is large, the e_g orbitals will not be occupied, and the t_{2g} orbitals may or may not be. The valence electron count is then somewhere between twelve and eighteen. Some examples of such complexes include: d^1 [Ti(en)$_3$]$^{3+}$ which has thirteen valence electrons; d^2 [Re(NCS)$_6$]$^-$ which has fourteen valence electrons; and d^4 [Os(SO$_3$)$_6$]$^{8-}$ which has sixteen valence electrons. If, however, Δ is small it will not be too energetically expensive to occupy the e_g orbitals and so they will be occupied if it means that another bond can be formed. In such cases the valence electron count could be as high as twenty-two, for example d^9 [Cu(H$_2$O)$_6$]$^{2+}$ has twenty-one valence electrons.

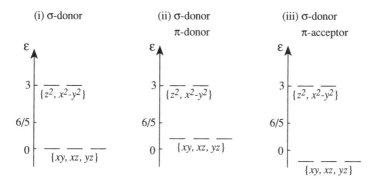

Fig. 5.6 Schematic illustration of d orbital energy level diagrams in an octahedral field due to (i) six σ-only ligands, (ii) six σ-donor and π-donor ligands, and (iii) six σ-donor and π-acceptor ligands.

π-donor ligands cause the t_{2g} d orbitals (which are non-bonding with σ-only L) to become anti-bonding. This makes less than eighteen-electrons an attractive option. On the other hand, π-acceptor ligands stabilise these same orbitals so favour their full occupation. If the L are both strong σ-donors, so making e_g unfavourable, and strong π-acceptors, making t_{2g} favourable, then precisely eighteen-electrons will be favoured. Some examples include: d^4 Ti(cyclopentadienyl)$_2$(CO$_2$)$_4$; d^6 [Co(NH$_3$)$_6$]$^{3+}$; and d^7 MnH(CO)$_5$.

The Eighteen-Electron Rule and Non-Octahedral Systems

For geometries other than {6,0} the eighteen-electron rule may also be observed, but not necessarily for the same reasons as in the octahedral case. For example, tetrahedral complexes with strong π-acceptor ligands will have eighteen-electrons, since the ligand π system has orbitals of both e and t_2 symmetry (as we saw for SiF$_4$ in §2.2.1), so eight electrons will be in σ bonds and ten in d orbitals. Trigonal bipyramidal ML$_5$ complexes probably also favour eighteen-electrons about the metal under these circumstances as $d_{x^2-y^2}$ and d_{xy}, which are destabilised by σ interactions, are stabilised by π-acceptors. In general, the result will depend on the relative strengths of σ and π interactions.

By way of contrast, square planar *trans*-{6,-2} complexes with strong σ-donor ligands are almost always sixteen-electron species even when the ligands are also strong π-acceptors. This is because $d_{x^2-y^2}$ will be strongly destabilised and hence will remain unoccupied (cf. Fig. 5.4).

5.1.5 Steric Versus Electronic Effects on Transition Metal Complex Geometry

Although we shall not be able to resolve the steric versus electronic debate for transition metal complexes, we are now in a position at least to understand it. In general, the shape of a transition metal complex arises as a balance between steric and bonding interactions. Determining the geometry of transition metal complexes may generally be reduced to the need to consider three factors: (i) the short range L-L repulsions (which may, for conceptual simplicity, be viewed in terms of a hard sphere repulsion), (ii) the M-L bond lengths, and (iii) the longer range L-L attraction due to dispersive or electrostatic interactions. Each of these influences has an optimal shape and C_N, but they are seldom the same. After considering C_N and shape in general, the remainder of this section is devoted to case studies of *tris*-chelate systems, the *trans*-effect and *trans*-influence, and Jahn-Teller active systems. These last topics are cases where the interplay of electronic and steric factors is very strong.

Which Coordination Number?

The dominant steric factor is the repulsive interaction between the ligands when they are squashed too closely together; thus the size of L limits the maximum number of ligands that can fit around a metal ion. Generally, $C_N \leq 6$, though we have already seen some examples of transition metal complexes with $C_N = 7, 8$, and 9 (§5.1.1). On the other hand, bonding interactions are favourable, and so the valence electrons are used in a way that optimises the number of bonding interactions without generating significant antibonding character. The general conclusion to be drawn is that Groups 3 - 9 are dominated by C_N of six, but Groups 10 and 11 can show more variation in both C_N and geometry.

If Δ is small then, as discussed above, the d orbitals are largely non-bonding in character. This means the metal valence electron count is not crucial and values in the range 12 - 22 can be supported. Thus, it is probably most favourable to maximise the number of M-L bonds. In such a case it is then the L-L interactions that determine C_N and a value of six is most likely, that being the close-packed arrangement consistent with the size of most transition metals and ligands. However, size is not the only factor in L-L interactions. For example, halogens are electronegative elements and so tend to be negatively charged within a complex. When this happens, the L-L repulsion is greatly enhanced, and more spacious geometries than octahedral may result; thus tetrahedral $[ML_4]^{2-}$ geometries are common when L is a for halogen.

If Δ is large, then for metals with electronic configurations up to d^6 an electron count of less than or equal to eighteen and a C_N of six are expected. For d^7, d^8, and d^9 systems, however, O_h complexes have the antibonding e_g d orbitals occupied. Such systems often adopt {5,0} or {6,-1} ML_5 geometries to avoid this, though ML_6 is still more common; d^8 and d^9 may also adopt ML_4 geometries. If electron count and steric factors complement one another the ideal scenario is {6,0}, {5,0} or {4,0} eighteen-electron systems. However, strong σ bonding can result in a particularly large destabilisation of the {5,0} $d_{x^2-y^2}$ orbital, and so if the {6,0} structure is not viable, d^8 systems will adopt a *trans*-{6,-2} with sixteen electrons instead, as discussed above.

O_h d^6 systems represent a situation in which the L-L interactions and the electronic effects coincide: for most ligands, the {6,0} template is a close-packed arrangement that will maximise any L-L attraction, while the O_h eighteen-electron structure allows full occupancy of bonding *mo's* yet leaves all the antibonding *mo's* unoccupied. In this way it is possible to understand the great stability observed for d^6 ML_6 complexes in comparison with many other transition metal complexes. A good example of this stability is provided by the low spin $[Co(1,10\text{-phenanthroline})_3]^{3+}$ which can be resolved into its chiral enantiomers and left in solution for months. The corresponding Co^{II} complex is d^7 and racemizes readily. As electron transfer between $[Co(1,10\text{-phenanthroline})_3]^{2+}$ and $[Co(1,10\text{-phenanthroline})_3]^{3+}$ is rapid, even a very small amount of $[Co(1,10\text{-phenanthroline})_3]^{2+}$ in a solution of $[Co(1,10\text{-phenanthroline})_3]^{3+}$ will catalyse racemization of the latter complex.

Which Template?

Determining C_N often specifies the template for the molecular geometry as well. This is particularly true of the ML_6 complexes, which are almost always octahedral, though perhaps with some degree of distortion as for *tris*-chelate complexes $M(LL)_3$ (see below). C_N's of five or four, however, both give rise to two common distinct templates: {5,0} or {6,-1}, and {4,0} or {6,-2} respectively. Which of these is observed is usually determined by a complicated interplay of steric and electronic factors as illustrated by the C_N's, electron counts and geometries of the d^8 Ni(II)-Pd(II)-Pt(II) and d^9 Cu(II)-Ag(II)-Au(II) triads, which are perhaps the most varied of the transition metal series.

Table 5.1 Some Group 10 and 11 geometries.

Complex	Geometry	Complex	Geometry
$NiCl_4^{2-}$	{4,0}	$Pd/Pt(CN)_4^{2-}$	{6,-2}
$NiBr_4^{2-}$	{4,0}	$Pd/PtenCl_2$	{6,-2}
NiI_4^{2-}	{4,0}	$Cu(CN)_4^{3-}$	{4,0}
$Ni(CN)_4^{2-}$	{6,-2}	$Cu[SC(NH_2)(CH_3)]^{4+}$	{4,0}
$NiCl_2(PMe_3)_2$	{6,-2}	$Ag[SC(NH_2)(CH_3)]^{4+}$	{4,0}
$NiCl_2(PPh_3)_2$	{4,0}	$Au[SC(NH_2)(CH_3)]^{4+}$	perhaps {6,-2}
$NiCl_2(PHPh_2)_3$	{5,0}	$(NH_4)_2CuCl_4$	{6,-2}
$NiBr_2(PEtPh_2)_2$	{4,0}, {6,-2}	Cs_2CuBr_4	distorted {4,0}
$NiBr_2(PPh_3)_2$	{4,0}	$[Ag(py)_4]^{2+}$	{6,-2}
$NiI_2(PMePh_2)_2$	{4,0}	$[Ag(bipy)_4]^{2+}$	{6,-2}
$Pd/Pt(NH_3)_4^{2+}$	{6,-2}	$[AuCl_4]^-$	{6,-2}
$Pd/PtCl_4^{2-}$	{6,-2}		

d^8 Ni is often in equilibrium between {4,0}, {5,-1}, *trans*-{6,-2} and {6,0}. The first three of these are sixteen-electron geoemtries, though the {6,-2} geometry is usually regarded as having coordinating solvent molecules as well, while the last is a twenty-electron system. Ligand size also has a strong influence on the geometry. Increasing ligand size favours {4,0} in Ni(II) complexes, *e.g.* alkyl phosphines favour square planar geometries but the bulkier aryl phosphines result in tetrahedral geometries, the change-over point being $Ni(PRPh_2)_2^{2+}$ (Table 5.1). $NiX_2(PR_3)_2$, X = Cl, Br, I, show analogous behaviour, the Cl compounds being planar, the I compounds being tetrahedral and the Br compounds being both or either geometry. Similarly, $NiX_2(LL)$ and $Ni(LL)_2$, where LL is a bidentate ligand and X = Cl, Br, I, NCS or NCO, also often adopt both planar and tetrahedral geometries. The length of the bite of the bidentate ligand can be a determining factor here. The effects of increasing M-L bond length in ML_4 down a triad is seen if L and C_N are held constant: Ag, Au, Pd and Pt complexes are invariably planar, whereas both square planar and tetrahedral Ni complexes are found, and Cu complexes are usually tetrahedral. In summary, d^8 Ni favours a square planar geometry if there is no steric crowding, but may also adopt an octahedral geometry to maximise the bond energy, whereas ,due to the decrease in bond strengths down the periodic table, Pd and Pt are never {6,0}. In the context of M-L bond strain it should also be noted that the bonds in square planar Ni(II) are shorter than in tetrahedral Ni(II) by about 5% due to the difference between high spin and low spin electron configurations (see §5.1.3).

Steric Interactions in Tris-Chelate Complexes

There are a number of approaches to transition metal complex geometry that work entirely with steric factors and ignore the details of electronic structure almost completely. These models were no doubt motivated by the difficulty in carrying out quantitative calculations: even with modern computer power there are still too many electrons in such systems to allow MO calculations on transition metal complexes to be routine. Such approaches have been surprisingly successful, which in itself is proof that electronic effects are not always the most important factor in determining metal complex geometry, especially once C_N has been established. The non-bonded radii approach and the AAIM, which were discussed in §1.3.2, fall under the general heading of steric theories. Molecular mechanics includes some electronic aspects since it parametrises bond strengths and angles, however, it also ignores many of the subtleties of bonding and yet is proving very successful for many transition metal systems, as noted in §1.3.2.

Perhaps the simplest successful example of a steric theory is the ligand-ligand repulsion model of Kepert [12,13] in which the geometry of a transition metal complex, and in particular of *tris*-chelate complexes, is considered to result solely from the repulsion between ligating atoms. Kepert quantified this by adopting a functional form to describe how these repulsions varied with the distance between the ligands. As we pointed out in §1.3.2 in the context of the AAIM, it is usual to identify two sources of repulsion: orbital overlap and electrostatic interactions between species with the same charge. These may each be quantified through an r^{-n} function where r is the distance between the two species and n takes the values of ~12 and 1, respectively. In order to simplify its implementation, Kepert combined these into an "average" value of $n = 6$ (although it should be noted that the justification for this is purely convenience and not theoretical; since $n = 6$ is actually appropriate for the *attractive* dispersion interaction).

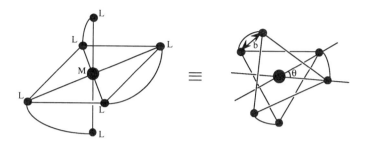

Fig. 5.7 Geometry of Λ-*tris*-chelate complexes.

Tris-chelate complexes, $M(LL)_3$, make an ideal case study because their distortion from octahedral geometry can be quantified in terms of just two parameters: the twist about the three-fold axis, θ, and the distance between the two ligating atoms of a single chelate (*i.e.* the "chelate bite"), b (Fig. 5.7). There are other characteristic L-L distances for these complexes, but they can all be expressed in terms of b, θ and the M-L bond length. We follow Kepert [12] and take b to be constant for a given ligand and adopt units such that the M-L bond length is one.

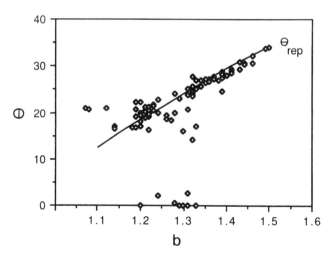

Fig. 5.8 (θ, b) values for *tris*-chelate metal complexes. Experimental data is summarised in reference [12]. The solid line is the twist angle, θ, which minimises an r^{-12} repulsive L-L interaction.

If there were no L-L attractive interactions, or if the L were close enough that the repulsive interactions dominated, then the geometry of $M(LL)_3$ would be that which minimises the repulsion; this is precisely the situation exploited by

Kepert. Now since we assume b is determined by the bidentate ligand, the only variable is θ and so the molecule will adopt the twist angle θ which minimises the non-bonded L-L repulsion. The solid line in Fig. 5.8 indicates the twist angle that minimises the repulsive L-L interaction for different b. Those with $b < \sqrt{2}$ have a geometry twisted from a regular octahedron towards a trigonal prism, while those with $b > \sqrt{2}$ twist further from the trigonal prismatic geometry. The diamonds correspond to experimental geometries. There is good qualitative agreement between the experimental points and the solid line. However, a purely repulsive interaction is not sufficient to account for the exact geometry for most complexes and for some, notably the D_{3h} {9,-3} systems, it seems to bear no relationship to what is found experimentally. The agreement is best for complexes with large ligands as might be expected for a purely repulsive potential.

For smaller ligands, $b < \sqrt{2}$, any L-L attractive interactions (cf. AAIM, §1.3.2) will mitigate the repulsive forces, and so lead to deviations from the geometries determined on the basis of repulsion alone. Indeed, the repulsion-determined angle, θ_{rep}, now becomes a transition state and a twist in either direction will stabilise the complex. The direction and amount of twist depends on the values of both b and the size of L, but can be quantified by adding an r^{-6} attractive term to the L-L interaction energy. Now, since the ratio of the attractive and repulsive coefficients, c_6/c_{12} (see Eq. 1.4), is a measure of the size of the ligand, it is convenient to compare different minimum-energy geometries for a given c_6/c_{12} and b. Some sample calculation results are given in Table 5.2 for $b = 1.1, 1.2,$ and 1.3.

Table 5.2 Some examples of geometry and energy parameters and the resulting AAIM L-L interaction energies for different *tris*-chelate geometries.

b	θ_{rep}	c_6/c_{12}	θ	$E(b,\theta)$
1.1	16.5	0.32	0	-0.063
			22	-0.067
		0.34	9.5	-0.068
			22.5	-0.071
1.2	21.3	0.36	0	-0.079
			14	-0.079
			24.5	-0.080
		0.40	17.5	-0.090
			25.2	-0.092
1.3	24.5	0.44	0	-0.117
			23	-0.112
			28.5	-0.115
		0.40	13.5	-0.101
			21.5	-0.101
			27.5	-0.102

For very small b values, e.g. $b = 1.1$, L must also be small and hence gives larger c_6/c_{12} ratios. The energy difference for this case between the stable more- and less-twisted relative to the θ_{rep} geometry is 4-6%, with θ_+ (more twisted) being more stable.

For intermediate b values, e.g. $b = 1.3$, the difference in energy between the θ_+ and θ_- geometries is only 1-2%. If L is small, θ_- is 2% more stable than θ_+, whereas for larger L, θ_+ is the more stable, albeit only slightly. Thus, intermediate size b with small L (large c_6/c_{12}) favour a geometry less-twisted from the repulsion-determined geometry. Intermediate b with intermediate L adopt geometries close to the repulsion-determined one. Intermediate b with large L result in some that are less-twisted and some that are more twisted than the repulsion-determined geometry. These conclusions are consistent with the data in Fig. 5.8.

Perhaps the most surprising feature of the geometries adopted by intermediate size ligands is that when L-L attractions lead to untwisting, the resulting geometry is often a long way from the repulsion-determined geometry; in some cases this can go to extremes and a geometry very close to a trigonal prism is adopted. As Table 5.2 shows there is in fact very little energy difference between twisted and untwisted forms, so factors other than L-L interactions could cause θ_- to be observed rather than θ_+ (or conversely). In particular, interactions involving the non-chelating atoms / groups in the ligands may have an effect, since they will be brought closer together in the trigonal prismatic form. Environmental factors such as crystal packing forces may also become significant when the L-L interactions give rise to only a weak θ dependence of the energy. Thus, e.g. $[Co(ethylenediamine)_3]^{3+}$ adopts θ values ranging about θ_{rep} depending on the counterion in the crystal.

Some detailed applications of the AAIM to these systems are given in reference [14].

Racemization Reactions of Tris-Chelate Complexes

One of the advantages of developing a simple but quantitative model for energetics along the lines of the last subsection is that it can also be applied to rearrangement reactions. *Tris*-chelate complexes are chiral and many of them can be resolved into their two mirror image enantiomers (usually denoted Δ and Λ - Λ was shown in Fig. 5.7) and isolated. Some will racemize subsequently if left standing in solution, leaving a mixture with no net chirality even though each individual molecule is either Δ or Λ. $[Fe(1,10-phenanthroline)_3]^{2+}$ is an example of this, having a half life (*i.e.* time for half the molecules to racemize) of about twenty minutes at room temperature in aqueous solutions.

In §2.3.3 we examined the symmetry allowed non-dissociative (*i.e.* non-bond breaking) enantiomerization reactions of *tris*-chelate complexes. The "push through" mechanism (Fig. 2.18a) has a planar hexagonal transition state that requires an M-L bond length stretch of about 40% if the bite angle can be squashed, otherwise it is even larger. The "cross over" mechanism (Fig 2.18b) requires an increase in M-L bond length of about 30% (again assuming the bite angle can be squashed). Even though M-L bonds are weak in chemical terms, these are still large extensions. Consequently, both mechanisms will involve high activation energies, and so will not be observed in practice.

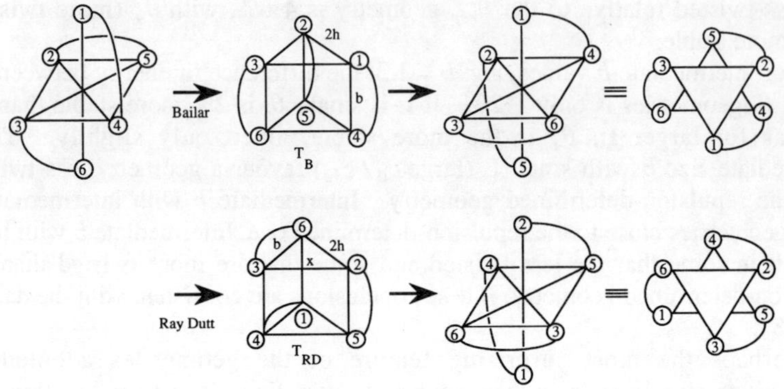

Fig. 5.9 Bailar and Ray-Dutt isomerization mechanisms for *tris*-chelate complexes.

The other two mechanisms, known respectively as the Bailar and Ray-Dutt twists (Fig. 5.9), have been drawn so that the underlying similarity of the mechanisms is apparent. The Ray-Dutt twist is usually illustrated in a completely different manner, but when viewed in this way it becomes clear that these two mechanisms involve similar twists about different axes (*cf.* §2.3.3).[15] The more favourable of the Bailar and Ray-Dutt twists will be that with the more stable transition state. Electronic factors are almost identical for the two, so let us examine the steric differences. It is desirable for there to be as little bond stretch as possible, so the Bailar and Ray-Dutt transition state geometries, T_B and T_{RD} respectively, will have M-L bond lengths as short as the L-L packing allows. For T_B, the position of M relative to the L is defined by the D_{3h} symmetry of the structure so that the M-L bond length becomes

$$d_{M-L} = (3h^2 + b^2)^{1/2}/2$$

where h is the non-bonded radius of L (so $2h$ is the distance between neighbouring L that do not belong to the same chelate, and is usually comparable in magnitude to b). In contrast, symmetry does not define the position of M along the two-fold rotation axis of T_{RD}; it has two M-L bond lengths of

$$(x^2 + b^2/4)^{1/2}$$

and four of

$$(h^2 + x^2 + a^2 - 2x(a^2 - b^2/4)^{1/2})^{1/2}$$

where x is the perpendicular distance between M and the line connecting ligands 2 and 5 in Fig. 5.9 and

$$a^2 = 4h^2 - (h - b/2)^2$$

For any b/h ratio the relative stabilities of T_B and T_{RD} may be compared by plotting M-L bond lengths as a function of x (the T_B bond length is, of course, independent of x) as in Fig. 5.10. Note that the activation energy is a combination of the strain energy for all six M-L bonds, but this will be dominated by the strain associated with the longest bond lengths. Inspection of Fig. 5.10 indicates that complexes with a small b value are more likely to proceed *via* the Bailar twist as the T_B M-L bond length is smaller than the average T_{RD} bond length for all x. For large b the Ray-Dutt twist becomes preferable, since values of x exist for which all the T_{RD} M-L bond lengths are smaller than those for T_B. At intermediate values of b both transition states will be of similar energy.

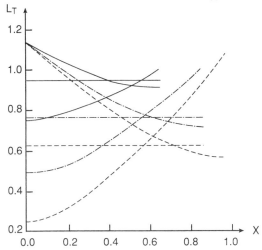

Fig. 5.10 Plots of M-L bond lengths at the transition state for *tris*-chelate isomerisations: $b/h = \sqrt{2}$, (---); $b/h = 2$, (-); $b/h = 2\sqrt{2}$, (---). Units are defined so that the L hard sphere radius, $h=1$ (instead of the M-L bond length, as used above) = 1. Straight thick lines are the Bailar twist transition state bond lengths; curved lines are the two Ray-Dutt twist transition state bond lengths. $b/h = 2$ is for chelate bite equalling the other L-L distances.

The Trans-Effect and the Trans-Influence

The interactions between ligands that are *trans* to each other in d^8 square planar Pt and Pd complexes can modify both kinetic and thermodynamic properties of the complex, and are usually referred to as the *trans*-effect and the *trans*-influence, respectively. The former describes the effect of a coordinated ligand upon the rate of substitution of ligands *trans* to it.[16] The *trans* influence, on the other hand, refers to the influence of a ligand upon structural properties such as M-L bond length, M-L vibrational frequency, and NMR coupling constants of a ligand *trans* to it.[17] Both features result from an interplay of steric and electronic factors and serve to illustrate some of the points discussed above.

We may understand the *trans*-influence by noting that bonds in complexes based on an octahedral template ({6,0}, {6,-1}, {6,-2} *etc.*) arise from overlap with metal orbitals that are either symmetric or antisymmetric to inversion; this means that the symmetry of the system (§2.2) ensures that the metal orbitals combine *trans* pairs of ligand orbitals into a single *mo*. Thus in a *trans*-{6,-2} D_{4h} complex, if a ligand along the $+x$ axis makes a particularly strong σ bonding interaction with the metal p_x, d_{z^2}, or $d_{x^2-y^2}$ orbitals, then it weakens the ability of the metal to use the same orbital in bonding with the ligand *trans* to it. Similarly, a good π acceptor ligand located at $+x$ bonds to M *via* the occupied (in d^8) d_{xz} or d_{xy} orbitals and weakens the bond of the ligand *trans* to it. Hence good σ donor and π acceptor ligands have strong *trans*-influences and the *trans*-influence series bears a strong resemblance to the spectrochemical series.

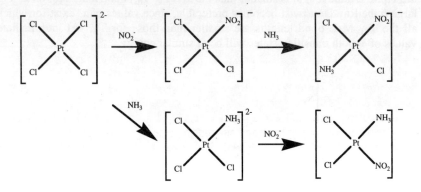

Fig. 5.11 Example of the *trans*-effect.

Being kinetic in nature, the *trans*-effect is more an issue of transition states than of stable molecular geometries. It describes the ability of ligands to labilize the substitution of the ligand opposite them. Despite this difference from the *trans*-influence, the *trans* effect also seems to be favoured by good σ donor and π acceptor ligands. It is illustrated by the two substitution reactions of Pt complexes shown in Fig. 5.11. The ligand *trans* to a Cl⁻ is always replaced.

There has been extensive discussion[18] about the rationale for the effect. Substitution reactions of square planar complexes almost certainly proceed *via* a five-coordinate square based pyramid {6,-1} that rearranges to a {5,0} transition state *via* a mechanism like mechanism β of Fig. 2.17, with atom 4 being the hole. The particular case for square planar complexes is illustrated in Fig. 5.12 with T (the *trans* ligand), D (the departing ligand) and A (the arriving ligand). The orientation of Fig. 5.12 obscures the equivalence of (D, T) and (L, L).

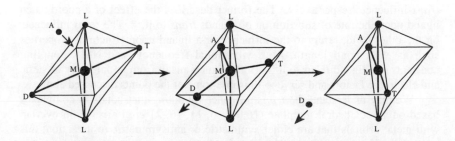

Fig. 5.12 Probable mechanism for substitution reaction reactions of {6,-2} complexes.

For the departure of D rather than an L to be favoured, we need to know: (i) why in {5,0} both T and D rather than an L are in the equatorial plane, and (ii) what feature of T stabilises {5,0} relative to the reactant thus making it a good *trans* effect ligand. In answering the first question we note that T or D are paired together by symmetry in the same *mo's* as discussed above, and so the symmetry of the reaction vibration will ensure that either both, or neither, are in the equatorial plane. Now, consider the approach of A to the square planar complex: it is a nucleophile, and so during approach to the complex it will avoid regions where there is significant π-donation of electron density from the metal to a ligand. Thus, if T is a strong π-acceptor ligand, A will approach on the other side of the complex, *i.e.* on the same side as D. Hence the A-M-T bond angle is bent towards the 120° angle characteristic of the equatorial plane. The reaction vibration is encouraged to continue in this direction because the equatorial position for T has good π-overlap opportunities with the metal (Fig. 5.5) relative to an axial position, and T is a good π-acceptor; this will also remove electron density from the *x-y* plane and so reduces the unfavourable σ antibonding interactions of the other equatorial ligands (Fig. 5.4). A further factor that encourages T to be equatorial is that D is almost certainly a weaker σ-donor (else it would not leave) than the *cis* ligands which will adopt the axial positions that favour stronger σ donors (Fig. 5.4).

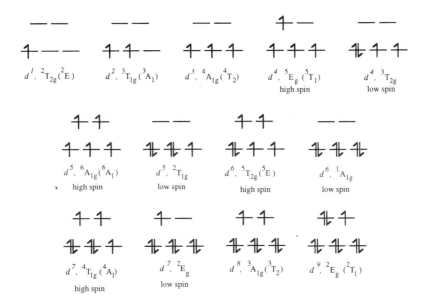

Fig. 5.13 Electronic ground states for O_h d^n systems, $n = 1$-9. T_d ground states follow from inverting the ordering of the orbitals; the ground state symmetry label for T_d is shown in parentheses, though it should be noted that T_d systems will always be high spin (*cf.* §5.1.3).

The answer to question (ii) requires consideration of the relative properties of the {4,0} reactant and {5,0} transition state. As discussed above, the properties of {4,0} will be modified by the *trans* influence, and so a good σ-donor and π-acceptor T will strengthen the M-T bond in the reactant at the expense of M-D. (The precise balance between M-D and M-T bond energies will depend on the σ and π character of D relative to T. For example, if D has no π character, then the π stabilisation of T will not affect the M-D bond.) However, in {5,0} these two ligands are no longer *trans* to one another, and so their bond energies are less interdependent. Thus, if T has good σ-donor character it makes a comparatively more stable bond in {5,0} than in {4,0}. This relative stabilisation of {5,0} will be strengthened further if T has any π-acceptor character. Good π-acceptor quality is more important for the *trans* effect than for the *trans* influence.

The Jahn-Teller Effect

The Jahn-Teller effect is the physical manifestation of the Jahn-Teller theorem [19] which states that: *stability and electronic degeneracy are mutually exclusive in non-linear molecules* (note that the electronic degeneracy refers to orbital degeneracy and not to spin degeneracy). The theorem thus applies to non-linear molecules whose wavefunctions would be doubly degenerate, E, or triply degenerate, T, if they adopted the high symmetry geometry.† Instead, the molecule distorts to take the wavefunctions to the A or B symmetry rows of the character table of a lower symmetry system. In this chapter we have so far considered only the symmetry of individual *mo's* (which describe the behaviour of single electrons); however in order to understand the Jahn-Teller effect we need to examine the symmetry of the wavefunction for the molecule as a whole. As discussed in §2.2.3, for transition metals this means we usually need consider only the metal d orbitals. Thus, determining the spatial degeneracy becomes a question of deciding how many ways there are to arrange the d electrons in the available d orbitals. The ground states for O_h and T_d are shown in Fig. 5.13. For O_h systems d^1, d^2, d^4, low spin d^5, high spin d^6, d^7 and d^9 are all electronically degenerate and therefore subject to the Jahn-Teller Theorem.

Although a proof of the Jahn-Teller theorem is not essential reading if one only wishes to use it, a knowledge of the proof is very helpful in understanding why the Jahn-Teller theorem works and what its consequences are and so a proof is given in Appendix 3. In the remainder of this section we will confine our attention to its use and consequences.

The main consequence of the Jahn-Teller theorem is that a non-linear molecule with a (spatially) degenerate ground state will distort away from the high symmetry geometry in order to break the degeneracy, and the type of distortions that can accomplish this can be determined by comparing the symmetry of the vibrations in the high symmetry geometry with the symmetry of the distorted form. Thus Cu^{II} d^9 complexes are often observed to be tetragonally distorted (Fig. 1.2), while the existence of square planar d^8 complexes can be thought of as an extreme version of this distortion. Note that a tetragonal distortion does not always lift the degeneracy. If the degeneracy lies in the t_{2g}

---

† The theorem also applies to five-fold degeneracy which is possible under icosahedral symmetry, but this symmetry is very rare in molecules.

orbitals rather than the e_g orbitals, the nature of the ligand determines whether a further distortion is still required. This is illustrated in Fig. 5.14 for d^6.

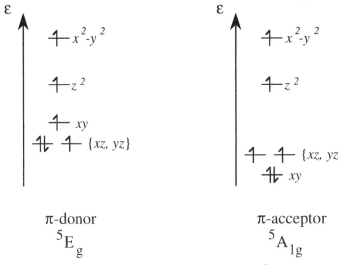

Fig. 5.14 Energy ordering of d orbitals of tetragonally distorted d^6 systems with π-donor and π-acceptor ligands, determined from Figs. 5.4 and 5.5.

It is also worth emphasising that, as with all symmetry results, *no* indication of magnitude is given by the Jahn-Teller Theorem. Indeed, often the distortion does occur but is undetectable, either due to the size of the distortion being too small or because the molecule alternates between different forms of the same fundamental distortion on a timescale that is too fast to be observed; an example of the latter arises in O_h complexes where the degeneracy can be removed by lengthening a *trans* pair of M-L bonds to form a tetragonal bipyramid, but this distortion could be achieved with any one of the three different *trans* pairs of ligands. Sometimes a much more subtle distortion is evident as in the case of the d^3 [V(H$_2$O)$_6$]$^{3+}$ systems discussed below.

The Geometry of Metal Hexaquo Systems

A number of crystal structures of these molecules are available from neutron scattering experiments. The alums CsFe(SO$_4$)$_2$.12H$_2$O and CsFe(SeO$_4$)$_2$.12H$_2$O show approximate T_h symmetry for the hydrogens (though a twist towards the D_{3d} structure occurs for the former and non-coplanarity of FeOH$_2$ occurs for the latter).[20] Best and Forsyth ascribed at least part of this to intermolecular hydrogen-bonding. In contrast, the d^2 [V^{III}(H$_2$O)$_6$][H$_5$O$_2$](CF$_3$SO$_3$)$_4$ adopts the all-horizontal D_{3d} structure of the [M(H$_2$O)$_6$]$^{3+}$ unit.[21] Cotton *et al.* attributed this geometry to the reduction of the O_h V(O$_6$)$^{3+}$ symmetry required by the Jahn-Teller effect for a d^2 ion. Distortion of the O_h V(O$_6$)$^{3+}$ to D_{3d} symmetry splits the triply degenerate t_{2g} orbitals into $a_{1g}+e_g$; for the all-vertical D_{3d} geometry the a_{1g} level is the more stable, whereas for the all-horizontal variant of this geometry a_{1g} is the less stable. Since the splitting is small, a high spin configuration is adopted, and so for d^2 the vertical D_{3d} geometry has a spatially degenerate ^3E_g

configuration, whereas the horizontal one has a $^3A_{2g}$ ground state which shows no spatial degeneracy (see Fig. 5.16). Cotton *et al.* predicted a vertical D_{3d} geometry for d^1 systems, while there should be no electronic driving force for d^3 systems and so we would expect the T_h geometry in this case.

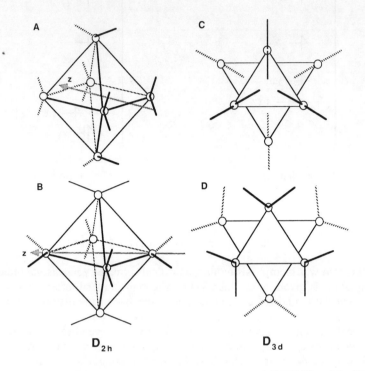

Fig. 5.15 Possible high symmetry orientations for hydrogens in $M(H_2O)_6$. A has D_{2h} symmetry, B has T_h symmetry and C and D have D_{3d} symmetry with the hydrogens all-vertical and all-horizontal, respectively. The dotted arrows indicate a two-fold axis in A and B.

5.2 Lanthanides and Actinides

Somewhat surprisingly, despite their valence orbitals including f in addition to s, p, and d, the geometries of lanthanide and actinide compounds are no more complicated to understand than those of the transition metals. In fact, the lanthanide and later actinide molecular geometries are perhaps the easiest in the periodic table to rationalise. The reason for this lies in the relative energies of their valence orbitals in the +2, +3, and +4 ions. Fig. 1.13 showed the periodic table drawn in an unconventional manner designed to emphasise where the lanthanides and actinides fit in.

Let us consider first the lanthanide row of the periodic table. The order of orbital occupancy in the neutral atoms is first $6s$ (Cs and Ba), then one $5d$ electron (La), then $4f$ (Ce to Lu). (The $5d$ and $5f$ orbitals are very close together in energy for the neutral atom so this assignment is a little arbitrary, but it makes for a clean story). However, this order is changed in the ions so that ionisation involves loss firstly of the 5d then the 6s electrons and then the 4f electrons. Thus, although the 4f orbitals have the highest energy in the neutral atoms, they become the lowest energy valence orbitals in the ions. In order to understand this inversion we note that the 6s and 5d electrons are very effective in shielding the 4f electrons from the nucleus (Fig. 1.14). Thus the removal of a 6s or 5d electron will considerably stabilise the occupied 4f orbitals in the ion. The net result is that the 4f electrons of lanthanide ions are buried within the core and play little part in the chemistry, in stark contrast to the d electrons and orbitals in transition metals.

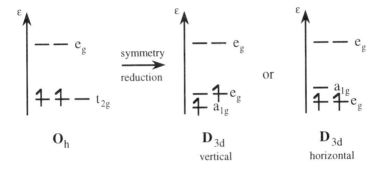

Fig. 5.16 Splitting of t_{2g} d orbitals of $M(H_2O)_6$ if MO_6 symmetry is reduced from O_h to D_{3d}.

One important consequence of these energetic factors is that the +3 ion (*i.e.* the ion formed after the loss of the 5d and 6s electrons) dominates the chemistry of lanthanide complexes. For the +3 ion, all valence electrons are 4f electrons and, since these are buried in the atomic core, electronic factors play only a minor role in determining the molecular geometry; steric-based models such as the AAIM should therefore work very well with lanthanide geometries. It should be noted, however, that this dominance of the +3 ion is not complete, and other oxidation states can be found. In particular, electron exchange interactions favour the formation of a high-spin f^7 state, resulting in quite extensive chemistry for Eu^{2+} (with promotion of a 6s or 5d electron to the seventh 4f orbital) and for Tb^{4+}.

In the case of the actinides, the shielding of the 5f electrons by the 7s and 6d is not as large as the corresponding shielding in the lanthanides. The Ac^{3+} ions of the early actinides (Th - Pu) have 5f electrons incompletely buried in the core. On the other hand, from about Cf onwards the poor shielding of one 5f electron by another coupled with the increased nuclear charge, are sufficient to cause the 5f electrons to be contracted into the core and lanthanide-like behaviour to re-emerge.[†] Even early in the actinide series, electronic effects are quite minor and they play only a small role in determining molecular geometry compared with the situation for transition metals. Thus, molecular shape arises largely from steric

---

[†] The contraction of electron density towards the core, and hence decrease in atomic size, in the actinides, lanthanides and also transition metals is often referred to as the "lanthanide contraction".

interactions. This is a good example of greater complexity actually leading to simpler behaviour in chemistry. The larger number of f orbitals and their more complex spatial distribution[†] compared with the d orbitals will admit many more electronically acceptable geometric arrangements of the complex, and the differences between the various "acceptable" geometries is likely to become less significant.

Table 5.3 Lanthanide molecules and ions indicating the range of C_N's and geometries adopted.

Formula	CN	Geometry [††]	Reference
Nd[N(Si(CH$_3$)$_3$)$_2$]$_3$	3	distorted {3,0}	26
LnO$_7$, Ln=Pr-Gd	7	{9,-2} and {8,-1}	22
Ln=La-Sm	7	{8,-1}	22
TbF$_7^{3-}$, ZrF$_7^{3-}$	7	perhaps {7,0}	22
Yb(CH$_3$COCH=COCH$_3$)$_3$H$_2$O	7	{9,-2}	22
NH$_4$Pr(TTA)$_4$.H$_2$O	8	{8,0}	25
Ce(IO$_3$)$_4$	8	{10,-2}	22
[Ln(H$_2$O)$_6$Cl$_2$]$^+$, Ln = Nd, Sm, Eu, Gd	8	{10,-2}	22
LnF$_4$, Ln=Ce,Tb	8	{10,-2}	22
LnX$_3$, Ln=Y, Sm-Lu & X=F, TbCl$_3$, SmBr$_3$, LaI$_3$	8	{9,-1}	22
[Ln(H$_2$O)$_9$]$^{3+}$, Ln = La, Pr, Nd, Sm, Gd, Dy	9	{9,0}	22
La(bipy)$_2$(NO$_3$)$_3$	10	{12,-2}	22
Th(NO$_3$)$_4$.5H$_2$O	11		23
La$_2$(SO$_4$)$_3$.9H$_2$O	12	approx. {12,0}	22
Mg$_3$Ce$_2$(NO$_3$)$_{12}$.24H$_2$O	12	approx. {12,0}	21
(NH$_4$)$_2$Ce(NO$_3$)$_6$	12	approx. {4,0}	21
CsU$_6$F$_2$S	9	{10,-1}	27
(NH$_4$)$_4$ThF$_8$	9	{10,-1}	28

[††] {12,-2} ≡ D_2 bicapped dodecahedron; {10,-1} ≡ monocapped square antiprism; {10,-2} ≡ square antiprism; {9,-2} ≡ monocapped trigonal prism; {9,-1} ≡ bicapped trigonal prism; {9,0} ≡ tricapped trigonal prism; {8,-1} ≡ monocapped octahedron; {7,0} ≡ pentagonal bipyramid; {3,0} ≡ trigonal pyramid, in this case distorted to have bond angles of 118°.

Having established that we can ignore the f electrons in determining molecular shape, what factors are left? For lanthanides and actinides the geometry is determined in part by the number of valence orbitals available for bonding (usually $s+p+d = 9$, and $f = 7-n/2$, where n is the number of f electrons) and in part by steric factors. As both the size of Ln^{3+} ions (La radius = 1.06Å, Lu radius = 0.84Å) and also the number of unoccupied available valence orbitals decreases across the series, the dominant factor is steric. In fact a reasonable

---

[†] A typical f orbital has eight lobes.

approximation to the C_N and geometry is achieved simply by considering the ionic model usually applied to Group 1/17 ionic solids, where the C_N of a metal is determined by close-packing the anion around the metal ion; this also determines the geometry. Thus, depending on M and L, C_N's in the range 3 - 12 are observed for lanthanides and actinides. Some deviations will occur where the size of L and the M-L bond length are incompatible with a close-packed geometry. In these cases the distortions can be understood in terms of L-L attraction (using the AAIM methodology of §1.3.2), and the geometry described by an n-vertex polyhedral template with h holes, $\{n,-h\}$ (§2.3.1). The direction of distortion will always be towards the next larger size polyhedral shape. The different templates are illutrated in Figs. 5.1-3 and 2.13-16.

Ligand geometries about lanthanide and actinide ions are available for a wide range systems. A small selection of these is given in Tables 5.3-4. In choosing these examples we have focused on systems in which the ligands are all similar chemical species. The available experimental data comes from both isolated molecules and ions, and crystal structures. While reiterating the warnings given in Chapter 1 about how the crystalline environment may perturb molecular geometry we note that, at least in the case of lanthanide or actinide halide ions, it is very likely that an isolated ion would remain associated with its counterion and, given the large number of ligands involved, the presence of the counterion is likely to cause a much bigger perturbation than does the crystal environment.

Table 5.4 Geometries of lanthanide and actinide halides in crystals. The labelling used refers to "standard" structures of the same form. L ≡ LaF$_3$; Y ≡ YF$_3$; P ≡ PuBr$_3$; U ≡ UCl$_3$; A ≡ AlCl$_3$; and B ≡ BiI$_3$. For a description of these structures see the text. Data from references [22, 24, 29].

	La	Ce	Pr	Nd	Pm	Sm	Eu	Gd	Tb	Dy	Ho	Er	Tm	Yb	Lu
MF$_3$	F	F	F	F		F/Y	F/Y	F/Y	F/Y	F/Y	P				
												Y	Y	Y	Y
MCl$_3$	U	U	U	U		U	U	U	P	P	P				
												Y	Y	Y	Y
MBr$_3$	U	U	U	P		P									
MI$_3$	P	P	P	P		B	B	B	B	B	B	B	B	B	B

	Ac	Th	Pa	U	Np	Pu	Am	Cm	Bk	Cf	Es	Fm	Md	No	Lr
MF$_3$	F		F	F	F	F	F		F/Y						
MCl$_3$	U		U	U	U	U	U	U	P/U	U					
MBr$_3$	U		U	P	P	P	P		A						
MI$_3$				P	P	P	B	B							

The labelling of lanthanide and actinide halide geometries in Table 5.4 is in terms of the following "standard" geometries seen in the crytsal structures. F ≡ LaF$_3$ has a nine coordinate molecular unit and is usually describes as a tricapped trigonal prism $\{9,0\}$ with seven short and two longer M-L bonds. F may also be describes as an *arachno*-{11,-2} structure (*cf.* Fig. 2.13) with the two of the four coordinate vertices being replaced by holes, the six-coordinate vertex being set at an optimal bond length, and the bottom left and right vertices being the long bonds. Y ≡ YF_3 is also a nine coordinate molecular unit usually described as a distorted trigonal prism with eight short and one long M-L bond. This is {10,-1}, with the hole being one of the apices. P ≡ $PuBr_3$ is an eight coordinate geometry with one much longer bond; however, in every other way it resembles U ≡ UCl_3 which is a regular tricapped trigonal prism so it has been suggested [24] that it is only crystal packing constraints that distort P. Thus, P ≈ U ≡ {9,0}. Finally, A = $AlCl_3$ and B = BiI_3 are {6,0}. Seven and eight coordination geometries are noticable by their absence, presumably because nine bonds using all *s*, *p* and *d* orbitals is energetically favourable and the M-L bond length increases required to add two L to {7,0} or 1 L to {8,0} for a given L-L distance is not significant (see Table 2.7).

References

(1) Bailar, J. C.; Emeleus, H. J.; Nyholm, R.; Trotman-Dickenson, A. F. *Comprehensive Inorganic Chemistry*; Pergamon Press: Oxford, 1973.
(2) Cotton, A. F.; Wilkinson, G. *Advanced Inorganic Chemistry;* 5th ed.; Wiley-Interscience: New York, 1988.
(3) Greenwood, N. N.; Earnshaw, A. *Chemistry of the Elements;* 1st ed.; Pergamon Press: Oxford, 1984.
(4) Huheey, J. E. *Inorganic Chemistry: Principles of Structure and Reactivity;* 3rd ed.; Harper International: New York, 1983.
(5) Porterfield, W. W. *Inorganic Chemistry: A Unified Approach*; Addison-Wesley Pub. Coy.: U.S.A., 1984.
(6) Griffith, J. S. *J. Inorg. Nucl. Chem.* **1956**, *2*, 229.
(7) Orgel, L. E. *J. Chem. Phys.* **1955**, *23*, 1819.
(8) Purcell, K. F.; Kotz, J. C. *Inorganic Chemistry;* Saunders: Philadelphia, 1977.
(9) Dunn, T. M. *Some Aspects of Crystal Field Theory*; Harper and Row: New York, 1965.
(10) Warren, K. D. **1977**, *16*, 2008.
(11) Glerup, J; Mønsted, O.; Schäffer, C. E. *Inorg. Chem.* **1976**, *15*, 1399.
(12) Kepert, D. L. *Progress in Inorganic Chemistry* **1979**, *23*, 1.
(13) Kepert, D. L. *Inorg. Chem.* **1972**, *11*, 1561.
(14) Rodger, A.; Johnson, B. F. G. *Inorganica Chimica Acta* **1988**, *146*, 37.
(15) Rodger, A. *Inorganica Chimica Acta* **1991**, *185*, 193-200.
(16) Basolo, F.; Pearson, R. G. *Prog. Inorg. Chem.* **1962**, *4*, 381.
(17) Langford, C. H.; Gray, H. B. *Ligand Substitution Processes*; W.A. Benjamin Inc.: 1965.
(18) Belluco, U. *Organometallic and Coordination Chemistry of Platinum*; Academic Press: London, 1974.
(19) Jahn, H. A.; Teller, E. *Proc. Roy. Soc.* **1937**, *161A*, 220.
(20) Best, S. P.; Forsyth, J. B. *J. Chem. Soc. Dalton* **1990**, 395-400.
(21) Cotton, F. A.; Fair, C. K.; Lewis, G. E.; Mott, G. N.; Ross, F. K.; Schultz, A. J.; Williams, J. M. *J. Amer. Chem. Soc.* **1984**, *106*, 5319.
(22) Moeller, T. *The Lanthanides*, 44, in *Comprehensive Inorganic Chemistry*, 4, Bailar, J.C. Jr, Emeléus, H.J., Nyholm, R.,Trotman-Dickenson, A.F. (eds), Pergamon Press: Oxford, 1973.
(23) Brown, D. *Actinide and Lanthanide Nitrates* in *Lanthanides and Actinides,* 7, MTP International Review of Science, Inorganic Chemistry (series 2) K.W. Bagnall (ed), Butterworths: London, 1975.

(24) Wells, A.F. *Structural Inorganic Chemistry 5th edn,*Clarendon Press: Oxford, 1984.
(25) Lalancette, R.A.; Cefola, M.; Hamilton, W.C.; La Placa,S.J. *Inorg. Chem.* **1967**, *6*, 2127-2129.
(26) Andersen, R.A.; Templeton, D. H.; Zalkin, A. *Inorg. Chem.* **1978**, *17*, 2317-2319.
(27) Brunton, G. *Acta Cryst. Sect. B* **1971**, *27*, 245.
(28) Ryan, R.R.; Penneman, R.A.; Rosenzweig,A. *Acta Cryst. Sect. B* **1969**, *25*, 1958.
(29) Brown, D. *Halides of the Lanthanides and Actinides*;John Wiley and Sons Ltd: London, 1968.

CHAPTER 6

Organometallic Compounds and Transition Metal Clusters

Contents

Introduction		139
6.1	Metal carbonyls	140
6.2	Transition metal clusters	142
	6.2.1 The metal polyhedron	143
	Electron counting schemes	144
	Free electrons on a sphere	146
	Polyhedral skeletal electron pair theory	146
	Cohesive Energy	147
	6.2.2 Metal polyhedron plus ligand polyhedron	149
	Clusters as a collection of (ML_j) fragments	149
	Clusters as a metal polyhedron encapsulated by a ligand polyhedron	150
6.3	Some examples	155
	$M_2(CO)_n$	155
	$M_3(CO)_n$ and fluxionality	155
	$M_4(CO)_{12-n}L_n$	157
	M_6L_n	159

Introduction

Organometallic chemistry is a rather loose label covering both the chemistry done by organic chemists where a metal is present in some apparently incidental role, such as being the template for the organic reaction, and the chemistry done by inorganic chemists usually working with transition metals and carbon based ligands. It should be noted, however, that the actual chemistry performed in both extremes is often the same, it is the aims, emphases and interpretations that differ. In general, the metal-ligand bonding of organometallic compounds involves the net sharing of electrons rather than the donation that was usually the case for the systems considered in the previous chapter, and carbon is most commonly the ligating atom; one main consequence of this sharing is that the metals are almost always in the zero oxidation state in organometallic compounds. However, no clear dividing line can be drawn; for example, halogens feature in both organometallic and non-organometallic systems, and ligands with phosphorous as the ligating atom are also considered the province of the organometallic chemist.

The emphasis in organometallic chemistry, at least from the inorganic side, is on transition metal cluster compounds which, by definition, have three or more metals bonded directly to one another. These also form the focus of this chapter though we shall begin with simple M(CO)$_n$ systems to establish many of the bonding principles appropriate for the larger systems. We shall also include molecules with only two metal atoms bonded directly to one another since, at least from a geometric point of view, they fit naturally with the larger molecules. Our aim is not to cover the breadth of organometallic chemistry, nor even to mention all the geometries that are possible since this area of chemistry has the most diverse range of molecular site geometries, as recourse to any general reference [1-4] will show. We shall try to establish some principles that determine the geometry adopted and, perhaps more importantly, present a framework for viewing cluster geometries that will be of assistance in visualising and remembering the structure determined for any specific compound that might be considered. A fairly complete, but not too complicated, view of the field of cluster chemistry is provided by the recent book of Mingos and Wales.[5]

6.1 Metal Carbonyls

Probably the most important ligand in organometallic chemistry is CO. It forms a bridge between the transition metal complexes of the previous chapter and the transition metal clusters that form most of the content of this chapter, and so it is useful to consider the transition metal complexes of CO. CO is also typical of many of the other ligands that play significant roles in organometallic chemistry. The MO energy level diagram of CO was shown in Fig. 1.20. In addition to the occupied orbitals shown in that figure it has low lying empty π* orbitals (*cf.* Fig. 1.18) that are ideal as π-acceptor orbitals as well as having occupied orbitals oriented in such a way as to be good σ donors. As discussed in §5.1.4, such ligands favour eighteen electrons about the metal to which they are bonded.

Thus CO requires empty metal orbitals for σ-donor bonding: usually *3d$_{z^2}$*, *3d$_{y^2-z^2}$*, *4s*, and *4p* for the first row transition metals and the corresponding orbitals for second and third row transition metals. For optimal bonding it also requires filled *d* orbitals - usually *3d$_{xy}$*, *3d$_{xz}$*, *3d$_{yz}$* - to provide electrons for "π back-bonding" contributions to the bonds. The metals are usually in low oxidation states so that the M electrons are not held too tightly and are available for π donation from M to L. The bonding of CO to transition metals is therefore optimised in the middle of the periodic table and with little net electron transfer so the M-C interaction is nearly an ideal covalent bond. The formal eight electron count for the C and the O is satisfied by a double bond between them; however, as a result of the π donation from the metal it has more bonding character than a normal double bond. The charge on the C is approximately $(0.09\pm0.05)e$ and on the O $(-0.12 \pm 0.05)e$, where *e* is the elementary unit of charge.[6]

The covalent nature of the M-C bonds also means that it has been possible to isolate low C$_N$ (coordination number) uncrowded metal carbonyl complexes in matrices[7,8] as well as the more common and more stable higher C$_N$ systems. Fig. 6.1 gives some examples. The low C$_N$ complexes provide a frustratingly good illustration of the interplay of steric and electronic factors in determining stable molecular geometries. For example, the three coordinate $M(CO)_3$, M = Cr, Fe, Ni and Cu (d^6, d^8, d^{10}, and d^1s^1,[†] respectively) seem to behave according to steric arguments. They take progressively more close-packed geometries as the metal gets smaller across the period: pyramidal-{6,-3} for Cr, pyramidal-{4,-1} for Fe, and planar-{3,0} for Ni and Cu geometries. Similarly, $M(CO)_3$ is pyramidal for M = Mo, whereas when M = Rh, Pd, Ag, Ta, and Pt the geometry is planar.

The situation for four-coordinate systems is not so clear-cut, and often it is electronic factors that seem to determine geometry. Thus for the geometries of $Cr(CO)_4$ (fourteen electrons, distorted {6,-2} or {5,-1}), $Fe(CO)_4$ (sixteen electrons, distorted {5,-1} or {4,0}), and $Co(CO)_4$ (seventeen electrons, distorted {4,0}), distortions seem to enhance the bonding interactions: an occupied orbital becomes lower in energy while an unoccupied one is raised. $Ni(CO)_4$, however, has eighteen electrons and is {4,0}.

$M(CO)_5$ geometries seem to correlate with electronic factors (cf. Fig. 5.4) with the d^8 metals (Mn^-, Fe, Ru and Os) and d^5 V being trigonal bipyramidal, whereas d^6 and d^7 metals are square pyramidal with the angle between the axial and equatorial ligands as follows: 94° for W, 95° for Cr^-, 90° for Mn, 91° for Mo, 95° for Mo^-, and 90° for Re. For $Fe(CO)_5$ the axial bond lengths are 2 pm shorter than the equatorial ones due to electronic effects: the d_{z^2} orbital is unoccupied. If steric factors were dominant the axial bond would be longer.

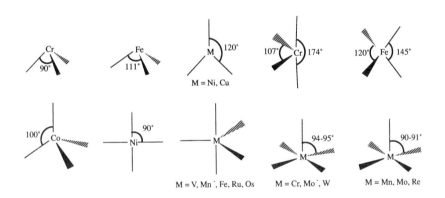

Fig. 6.1 Examples of metal carbonyl complexes.

The main lesson to learn is that when the system does not formally have eighteen valence electrons about the metal, beware of electronic factors causing sterically unexpected distortions. We shall see examples of this throughout the chapter. However, since the ligands of organometallic chemistry are typically σ-donor and π-acceptor ligands, they are likely to favour eighteen valence electrons about the metal. One corollary of the strong influence of the eighteen-electron rule is that if there is room, the system will take the number of ligands required to satisfy the electron count. Exceptions to the rule, such as $V(CO)_6$, occur when

---

[†] The energy orderings of the s and d orbitals is reversed in the presence of ligands from what it is in an isolated atom.

there is not enough room to accommodate another ligand. Thus no C_N is particularly favoured, in contrast to the predominance of octahedral geometries noted for transition metal complexes in the previous chapter. Furthermore, steric arguments of one kind or another have been applied quite successfully to the analysis of the geometries of clusters, as discussed below.

6.2 Transition Metal Clusters

Transition metal clusters, M_mL_n, have two (more strictly three) or more metal atoms which are directly bonded to one another, so not all metal electron density is available for M-L bonding. In addition, the L may attach terminally to one M or may be bridged between two or more M. The main differences between transition metal clusters and boranes (§3.3) are the involvement of d electrons in the bonding between cluster atoms, and hence a greater variety of orbitals available for bonding. As with main group clusters, the standard idea of a two-electron bond is not always applicable. In this chapter the difference between molecular structure and geometry becomes important for the way we shall approach the subject. Much of cluster chemistry revolves round X-ray crystallographic determination of newly synthesised compounds. Thus a great deal is known about bond lengths, bond angles and small variations in these. We shall ignore that body of data and look at clusters much more superficially, trying to get principles for determining and ways of describing the geometries that are observed.

The use of the different energy partitionings discussed below leads to different theoretical approaches to cluster bonding and has implications for the way cluster structures are represented. For smaller clusters this is only a trivial distinction, but for the larger ones it can make a big difference to how geometric information is portrayed. The traditional option, and the one generally adopted by the crystallographers, is to draw the metal polyhedron with edges linking nearest neighbour atoms, and then to draw bonds between each ligand and its nearest neighbour metal atom(s). In contrast to boranes, the M-M link that is bridged by a ligand is always illustrated. The alternative is to draw the metal polyhedron encapsulated in a polyhedron defined by the ligands. There are reasons for adopting either or both of these options. The former approach emphasises that the M-L bonds are the most important contributors to the bond energy, and the latter suggests that L-L interactions may be more significant in determining of geometry (see below). The former is more convenient for discussing substitution reactions and the latter for isomerization reactions. We shall use both representations in this chapter. Some of the possible bonding modes and cluster geometries are illustrated in Fig. 6.2 for molecules with two metal atoms. Other examples are given later in the chapter.

The basic problem with transition metal clusters from a theoretical standpoint is that they have too many electrons and atoms for any fundamental and rigorous theory. Thus it is necessary to make some reasonably drastic assumptions, although this does not mean that the resulting theories should be necessarily be discredited. The aim should be to view just enough of the complexity to be able to claim to understand transition metal cluster geometries. Predictions are made with the greatest of circumspection and even rationalising known results can be a dangerous past-time.

Having begun with such pessimistic words of caution we can now point out that some of the transition metal bonding theories have been very useful. Much of their success seems to be due to the fact that a lot of information is implied by the high symmetry of the systems being considered, and so a small amount of accurate energetic input data gives a lot of qualitative energetic output. The above warning, however, serves to remind us not to be surprised when nature, helped along by an inventive synthetic chemist, confounds any conclusions or generalisations we have made. Most approaches to transition metal cluster geometry focus first on the geometry of the M_m polyhedron and consider the ligands later (although sometimes this is done implicitly rather than explicitly). We shall also follow this route, though of course the bonding in the metal polyhedron cannot be completely separated from the requirements of the ligands.

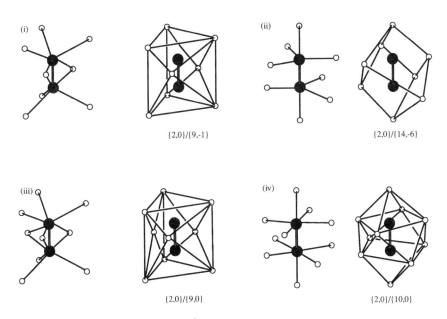

Fig. 6.2 (i) $Co_2(CO)_8$; (ii) $[Fe_2(CO)_8]^{2-}$; (iii) $Fe_2(CO)_9$; (iv) $Mn_2(CO)_{10}$. Metal atoms are represented by large black circles, and CO by smaller open circles. Each M-C-O bond angle is close to 180° for terminal CO ligands. The left-hand side picture of each pair emphasise the M-L bonds, while pictures on the right show the metal polyhedron and the ligand polyhedron. The labels given are according to §2.3.1 with the metal polyhedron given first and the ligand polyhedron second.

6.2.1 The Metal Polyhedron

It is tempting to say that each edge or nearest neighbour M-M link in the metal polyhedron corresponds to a chemical bond, but this is not necessarily the case for a cluster system. Although polyhedral edges are frequently considered to correspond to a *directional* bond, the bonding in clusters is such that parameterising the overall binding energy within this cluster system in terms of either edges or faces is usually unsatisfactory. Under certain circumstances it may be convenient to do so, but in reality bonds (considered as regions of high electron density) can not be localised on either edges or faces of the metal polyhedron. If one were to separate the components of the overall chemical glue holding the M_m unit together into edge bonds, then these bonds would nomally have bond orders of *less* than one. Take, for example, $[Co_6(CO)_{14}]^{2-}$ (Fig. 6.3). According to simple electron-counting theory (see below) this has fourteen electrons available for M-M bonding yet has an octahedral metal polyhedron; this means each of the twelve edges would have a bond order of 7/12 on average. We must therefore look at the metal polyhedron stability less simplistically. A number of approaches have been taken; the more successful and common ones are discussed below.

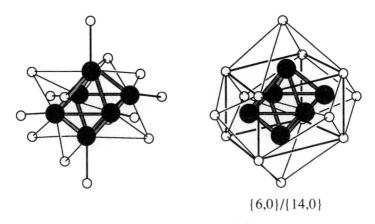

{6,0}/{14,0}

Fig. 6.3 $[Co_6(CO)_{14}]^{4-}$, depicted as *per* Fig. 6.2.

Electron Counting Schemes

Despite the many deficiencies of the various electron counting schemes they do form a helpful starting point for considering of transition metal cluster bonding and hence geometry - particularly that of the M_m central polyhedron. The different schemes are based on similar principles and the most widely used are Wade's electron counting rules, which are an extension of his rules for boranes (§3.3.1) to transition metals. For transition metals Wade assumed that three of the nine valence orbitals would be used for M-L bonding and three would be of sufficiently low energy to be consiered non-bonding. This leaves three orbitals per M for M-M bonding. The total M_m electron count is $n_e = \Sigma v + \Sigma x - 12m - c$, where v is the number of valence electrons on an isolated metal atom, x is the number of electrons contributed by each ligand, $12m$ electrons occupy ligand bonding or non-bonding orbitals, and c is the net charge on the cluster;[†] the Σ denotes a summation over all relevant orbitals / electrons. The electrons of any encapsulated ligand (*i.e.* within the metal polyhedron) are assumed to contribute directly to M-M bonding.

The shape of the M_m polyhedron is then taken to be an $(n_e - 2)/2$ vertex (triangulated) deltahedron. If $m = (n_e - 2)/2$ then the cluster adopts a *closo* metal polyhedron; if $m = (n_e - 4)/2$ it is *arachno*; etc. For example, $[Co_6(CO)_{14}]^{4-}$ (Fig. 6.3) has $n_e = 6 \times 9 + 14 \times 2 - 6 \times 12 + 4 = 14$. Thus M_m is predicted to be *closo*-{6,0}, as is in fact the case. The electron counting, at least in this simplest form, makes no suggestion about where the ligands are to be found. Similarly $Rh_6(CO)_{16}$ is predicted to have an octahedral metal polyhedron (Fig. 6.4). Problems arise for $Os_6(CO)_{18}$, however, which is predicted to be octahedral but is in fact a bicapped tetrahedron,[9] and for $[Os_6(CO)_{18}]^{2-}$ which is octahedral.[10] More discussion and examples of electron counting schemes can be found in reference [11].

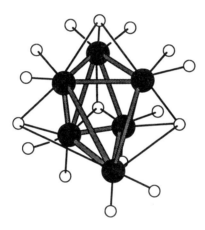

Fig. 6.4 Metal and ligand polyhedra for $Rh_6(CO)_{16}$.

In applying a simple electron counting scheme it should be noted that (i) not all ligands contribute two electrons; (ii) not all ligands are as consistent as CO in providing the same number of electrons in all possible ligating geometries (see *e.g.* reference[12] for electron donating properties of the more common organometallic ligands); (iii) metals on the right handside of the periodic table may be stable with an electron count of less than eighteen;[13] and (iv) as with boranes, the more holes the M_m polyhedron is predicted to have the less likely the electron counting answer is to agree with experiment, in part because more than three metal orbitals will be used in M-L bonding. For example, the dihydrido cluster $H_2Os_3(CO)_{10}$ has

---

[†] Note: c in this chapter is net charge on the cluster, whereas in Chapter 3 it was the net *negative* charge on the borane.

$n_e = 3 \times 8 + 10 \times 2 + 2 \times 1 - 3 \times 12 = 10$ so by Wade's rules as formulated above would require an octahedron with three vacant vertices and the twelve ligands held on by bonding through only nine metal orbitals. In such a case a simpler scheme involving twelve two-electron bonds to ligands and therefore four M-M bonds to satisfy the electron count on each metal seems more reasonable.

Free Electrons on a Sphere

Given the increasingly spherical appearance of the metal polyhedron as m increases it is not surprising that attempts were made to describe the M_m bonding in terms of free electrons on a sphere; however, there was only mixed success until Stone applied tensor surface harmonic (TSH) theory.[14] In his treatment he described σ, π and δ orbitals *via* spherical, vector, and tensor surface harmonics, respectively, thus including the differences between nodes due to the *ao* basis set, and nodes between atoms which related to anti-bonding character. Stone also ignored the structure of the cluster and assumed that, as far as the valence electrons are concerned, it can be treated as a perturbed spherical shell. The solutions to the angular part of the Schrödinger equation were then used to determine the *mo's* as linear combinations of the *ao's* with the coefficients determined by the magnitude of the spherical harmonic functions at the atomic sites. By assuming that the cluster orbitals retain the symmetry of the spherical system, that orbitals of different azimuthal (l) and magnetic (m_l) quantum numbers do not mix, and that obitals with the same value of l are degenerate, it is possible to derive an approximate MO diagram for any cluster. Two of the results from this approach are that: (i) m atoms require $m+1$ electron pairs for stable bonding as assumed by Wade's rules (except for $m = 4$), and (ii) triangular faces are the most stable. Stone and Wales later extended TSH to take the positions of atoms into account more explicitly.[15]

Ceulemans and Fowler[16] applied TSH to bonding patterns and electron counts of high symmetry transition metal clusters and compared the results with those of other models. They concluded that it was very simplistic but led to simple and transparent pictures of bonding; however, it neither predicts not rationalises the geometry adopted.

Polyhedral Skeletal Electron Pair Theory

Polyhedral Skeletal Electron Pair Theory was developed mainly by Wade and Mingos,[17-19] and may be viewed as an integration of all the electron-counting, electron-on-a-sphere, and empirical-MO theories. A detailed review of its development may be found in reference [13]. As recently summarised by Mingos and May,[20] the parts relevant for single transition metal clusters (rather than condensed or mixed transition metal - main group clusters) are as follows.

(i) Clusters with M's in a two-dimensional ring have $16m$ valence electrons. This is because each M requires an average total count of eighteen electrons and each one gains two electrons, one from each bond with its two neighbours.

(ii) Metal polyhedra with only three-connected M vertices have $15m$ valence electrons, as such systems have two-centre two-electron bonds.

(iii) Four-connected M_m polyhedra have $14m+2$ valence electrons as long as the M's lie approximately on a single spherical surface. This follows from TSH

theory and the fact that M(CO)$_k$ fragments generally use three M orbitals for skeletal bonding (so the M-M links are not two electron bonds).
(iv) Capped clusters require skeletal bonding *mo's* in addition to those of the parent.

Any electron counting scheme, such as the one given above, that is based on an eighteen valence electron count for a metal may have to be modified on the right hand side of the transition metal series. Further, clusters containing M(CO)$_4$ units often have unusual geometries as one valence orbital is significantly lower in energy than the others, so it is not always three orbitals that contribute to cluster bonding.[21]

Cohesive Energy

A completely different approach to the energy of the metal polyhedron was developed by beginning with the premise that M$_m$ was a fragment of bulk metal.[22] Transition metals have s, p and d electrons available for bonding; some of these will be involved with bonding to ligands and some to other metal atoms. If we remove the electrons (more strictly the electron density) required for M-L bonding from our initial consideration, the remaining s, p and d electrons bind the metal atoms together. Not all of the d electrons are involved in M-L bonding, so the M-M bonding might be expected to reflect the bonding characteristics of a small fragment of bulk transition metal. Woolley [23] developed this line in argument and wrote the energy of M$_m$ as

$$E_{M-M} = E_{rep}(s,p) + E_{attr}(d) \qquad (6.1)$$

$E_{rep}(s,p)$ is positive and due to the repulsion of valence s and p electrons by cores of adjacent transition metal atoms, it is short ranged. $E_{attr}(d)$ is negative and arises from the attraction of the metallic-type bonding of the d electrons. The form for the cohesive energy of bulk metal where each atom has less than ten d electrons has been shown by Woolley to be an adequate description of $E_{attr}(d)$ when account is taken of the reduced d electron density in a cluster caused by M-L bonding. Thus we can write the cohesive energy of the metal polyhedron of a cluster as:[24,25]

$$E_{M-M} \approx \Sigma (R_{M-M})^{-5} (Z_M)^{1/2} A \qquad (6.2)$$

where R_{M-M} is the nearest neighbour distance between metal atoms, Z_M is the number of nearest neighbours for M, *i.e.* the connectivity, A is a constant determined by the number of d electrons and the details of the atomic potential at each site, and the sum is performed over all nearest neighbours metal atoms. Broadly speaking, Z_M is a function of the polyhedral structure adopted by M$_m$; R_{M-M} reflects the size of the system; and A varies both as a function of the metal involved and of the ligand system, to the extent that the ligand system determines the number of electrons available for M-M bonding. The justification of this equation lies in the methods of solid state physics, such as chemical *pseudo-potential* and $X\alpha$ methods.[25] The success of Kepert's empirical parametrization of the atom-atom interactions in the core of a cluster in terms of an attractive and a repulsive inverse distance dependence (see below) supports this type of approach, showing that the interactions are comparatively short ranged.

For a given M_mL_n system, A is expected to be approximately constant for the different possible structures,.and so consideration of only the M_m cohesive energy would suggest that M_m will adopt the geometries with maximum cohesive energy, *i.e.* maximum numbers of nearest neighbours. This leads to polyhedra with the maximum number of triangular faces and maximum number of nearest neighbours, *viz.* tetrahedron for $m = 4$, trigonal bipyramid for $m = 5$, bicapped tetrahedron or octahedron for $m = 6$, tricapped tetrahedron or pentagonal bipyramidal for $m = 7$ *etc.*[26] For the larger systems, $m > 7$, perhaps one might expect to see variations in structure since cohesive energies vary so little (0.2% difference in cohesive energy between the pentagonal bipyramid and the capped octahedron). Johnson and Woolley concluded that clusters favour more compact close packed bonding arrangements than boranes due to the bonding effects of the d electrons contributing to the M-M cohesive energy.[27] As always there are exceptions, such as the Group 10 butterfly clusters.[28] In practice, a variety of M_m structures are observed for all m. This reflects the large (possibly dominant) contribution to the total energy made by the M-L interactions. A metal polyhedron which does not maximise the cohesive energy may be observed if this M_m structure gives rise to more favourable M-L interactions than does the most stable metal polyhedron.

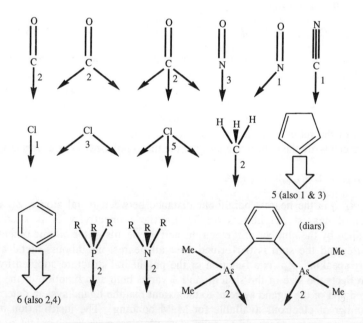

Fig. 6.5 Some common organometallic ligands. The arrows indicate the number and direction of bonds a ligand forms with M_m: one bond means a terminal ligand, two an edge bridged ligand (between two M's), and three a face capping ligand. The numbers indicate the number of electrons donated to M_m bonding.

6.2.2 Metal Polyhedron Plus Ligand Polyhedron

In main group systems the ligands are largely ignored, as they are usually hydrogens. By way of contrast, when discussing the geometry of transition metal clusters attention usually focuses on the ligands since M_m usually adopts $\{m,0\}$ and it is the arrangement of ligands that provides the variety, as shown by the figures in this chapter. The identity of the ligands affects M_m only in so far as different sized ligands distort the polyhedron to different extents, and different ligands donate different numbers of electrons to the cluster bonding. Fig. 6.5 illustrates some common organometallic ligands indicating how they bond and the number of electrons they donate to the cluster (*i.e.* to M_m) bonding.

Two types of approach to the ligand polyhedron follow from the different schematic ways the energy of the cluster may be divided up as discussed above. The cluster is seen either as a collection of ML_j fragments, or as a metal polyhedron enclosed in a ligand polyhedron and the energy may then be written symbolically as done below. Fig. 6.6 illustrates the difference schematically.

Clusters as a Collection of (ML_j) Fragments

When cluster compounds are studied after transition metal complexes, they seem to be collections of transition metal complexes with some ligands removed to enable M-M bonding. In order to make calculations feasible in the early 1970's, it was assumed that one could treat *e.g.* $M(CO)_n$, $n = 2,3,4$, $M(C_6H_6)$ and $M(C_5H_5)$ as independent fragments using MO methods,[29,30] and then join the fragments together. This approach is still the basis of accurate MO calculations. Thus, the most common approach to cluster geometries has been to consider building a cluster from ML_j fragments with the energy expressed as a sum of interaction energies between adjacent fragments *j* and *k*:

$$E_{total} = \Sigma\, E[(ML_j)\text{-}(ML_k)] \tag{6.3}$$

The M-L bonds are assumed to be already in place and the only question that needs to be addressed is whether the ML_j and ML_k may interact so as to produce a more stable product. The most successful qualitative way of answering this question has been the isolobal analogy whereby simpler molecular fragment systems whose "relevant" electronic structures are the same as those of the ML_k fragments of interest are found. The assumption is then made that the interactions between the ML_k fragments of interest are the same as those between the simpler fragments. In practice, an isolobal analogy has required that the frontier *mo's* (*i.e.* highest occupied and lowest unoccupied orbitals) of the fragments be similar in symmetry, extent in space, and energetics. The most common isolobal analogies are between ML_k fragments and boron hydride or organic subunits such as methyl or methylene groups; *e.g.* $M(CO)_3$ is isolobal with B-H.[31] The isolobal analogy has been used with success in a number of instances. For example, Shaik *et al.* were able to explain the bonding of bridging carbonyls by taking it to be isolobal with methylene,[33] and Hoffmann found that the bonding in clusters containing only terminal carbonyl ligands could be analysed using an isolobal analogy between CH_n^{2-} and $M(CO)_n$ fragments.[31] Early MO calculations on $[Co_6(CO)_{14}]^{4-}$

supported both the isolobal approach and also electron-counting for the metal polyhedron.[32]

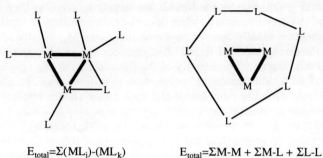

$E_{total}=\Sigma(ML_j)\text{-}(ML_k)$ $E_{total}=\Sigma M\text{-}M + \Sigma M\text{-}L + \Sigma L\text{-}L$

Fig. 6.6 Schematic representation of ways of dividing a cluster into sub-units.

It is interesting to note that, although based in an MO formalism, the isolobal analogy deals with localised orbitals rather than the completely delocalized orbitals of MO theory. The approach seems limited only by the accuracy of the analogy and so provides further support to the use of localised bonding approaches to understand molecular geometry.

At the present stage of application one might expect an isolobal analogy to give any symmetry determined features of the fragment interactions, and to reproduce other features to the extent which the interacting orbitals are in fact isolobal. Problems are therefore most likely to arise when the specific nature of d orbitals comes in to play in the transition metal clusters, or the greater variety of possible bonding orbitals becomes relevant (*i.e.* when non-frontier orbitals take part in the bonding). Woolley[34] has warned of the problems that may be encountered when the M-M bonding is investigated using isolobal analogies with main group units, due to the importance of d electrons in the M-M bonding and the fact that the main group fragments only involve s and p orbitals and so cannot reflect the behaviour of d orbitals exactly. A good illustration of this is provided by the work of Evans and Mingos[21] on $Os(CO)_4$ as a vertex fragment: CH_2 is often considered to be isolobal to $Os(CO)_4$; however, $Os(CO)_4$ has additional orbitals which may become involved in the bonding and enable it to bond to three osmiums, whereas CH_2 has no such flexibility.[31] In this instance it proved necessary to consider the available orbitals on the $Os(CO)_4$ fragment and determine its interaction with the rest of the cluster properly, not *via* the shortcut of an isolobal analogy with a more familiar system. Further, Au and to a lesser extent Pt clusters cannot be treated this way due to contributions to bonding from $6s$ orbitals.

Clusters as a Metal Polyhedron Encapsulated by a Ligand Polyhedron

The alternative energy partitioning is into M-M, M-L, and L-L contributions,

$$E_{total} = \Sigma E[M\text{-}M] + \Sigma E[M\text{-}L] + \Sigma E[L\text{-}L] \qquad (6.4)$$

and follows naturally from a combination of the cohesive energy descriptions of the metal polyhedron (§6.2.1) and the AAIM view of a transition metal complex

(§5.1.5). Instead of viewing the cluster as a "cluster" of transition metal complexes with a few ligands omitted, this partitioning is equivalent to viewing the cluster as a transition metal complex with the metal replaced by a "cluster" of metals and the ligands arranged about the enlarged (non-spherical) metal centre. Implicit in this description is the idea that the ligands provide electrons to the cluster bonding, usually so that each metal achieves an eighteen-electron count; yet the details of the M-L bonding are not important so long as reasonable M-L bonds are formed. This idea is not as heretical as it may sound and we have already given a number of examples of metal carbonyl systems in which the eighteen-electron count favours a variety of non-octahedral geometries. One consequence is that theories, such as Stone's, which treat the metal polyhedron as a sphere and ignore the details of the M-L bonding, are often very successful. The other difference between a simple transition metal complex and the enlarged cluster analogue is that, in addition to a range of LML bond angles, the ligands of a cluster may also be close enough to two or more metals to be considered as bonded to them all. The notation used to indicate this is μ-2 *etc.* The distribution of ligands about M_n is always fairly uniform to ensure an even electron distribution.

Thus, in the context of dividing cluster energies according to Eq. (6.4), we conclude that once the metal electrons involved in M-M and M-L bonding have been ascertained, the details of the ligand orientation about the metal polyhedron are largely determined by the smallest magnitude contribution to the energy: $\sum E[\text{L-L}]$. In other words, the versatility of the M-L bonding means that although the M-L bond energy is significant, it is seldom structure determining; and this is despite the fact that M-M bonds with bridging carbonyls have most of the electron density in the M-L links.[35] We note, however, that cluster chemistry is always rich in counter examples, and this is amply illustrated by the geometry of $[\text{Rh}_6\text{C(CO)}_{13}]^{2-}$ which is contrary to many of the general conclusions drawn in this chapter.[36] Its geometry is shown in Fig. 6.7 and will be considered in more detail below. Questions remain as to how the L-L interactions are optimised, and then how the metal polyhedron is oriented within the ligand polyhedron.

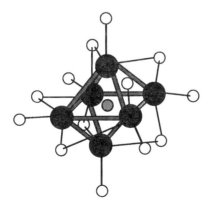

Fig. 6.7 $[\text{Rh}_6\text{C(CO)}_{13}]^{2-}$

Some work has been focused explicitly on the geometry of the ligand polyhedron. Following earlier studies on other systems, Benfield and Johnson[37,38] took the only effect of the M-L bonding interactions to be to define a spherical shell for the ligands, and then determined the ligand polyhedron by minimising ligand-ligand repulsion. They arbitrarily chose a repulsion exponent of -6 (as with the transition metal studies performed by Kepert (§5.1.3), this is inappropriate; however the results were not be greatly affected by the exponent since only repulsion forces were considered). They focused on systems with larger ligand polyhedra (eight or more ligands) and determined the ligand polyhedra that minimised repulsion. They developed a notation for describing ligand polyhedral geometries which relates to the way the polyhedron is drawn. We shall use a modified form of it here since it enables geometries different from those based on *closo* templates to be described. For this notation the polyhedron is divided into a series of parallel layers, and a number used to denote how many vertices are in each layer; parentheses are used to indicate if a set of vertices is aligned with respect to the first plane containing more than one vertex, square brackets to indicate if it is staggered, while no brackets are used for any other orientation. In this notation a bicapped square antiprism (Fig. 6.8) would be written: {(1):(4):[4]:(1)}. The numbering does depends on the orientation chosen for viewing the polyhedron, but is none-the-less a helpful guide. It is convenient to extend their notation to account for polyhedra with holes. A bicapped trigonal prism (Fig. 6.8) viewed down the three-fold axis of the rectangular prism is {(3):[3,-1]:(3)}. In this way Benfield and Johnson determined an ordering of favourability for ligand polyhedra of any specified size. It is as follows:

L_8: square antiprism {(4):[4]}; dodecahedron {(1):(4):(3)}; {(1):(5):(2)}; cube {(4):(4)}; and bicapped trigonal prism {(3):[3,-1]:(3)}.

L_9: tricapped trigonal prism {(3):[3]:(3)}; monocapped square antiprism {(1):(4):[4]}; {(1):(5):(3)}; and monocapped cube {(1):(4):(4)}.

L_{10}: bicapped square antiprism {(1):(4):[4]:(1)}; {(1):(3):[3]:(3)}; {(2):(4):(2): [2]}; {(2):(4):[4]}; {(1):(6):(3)}.

L_{12}: icosahedron {(1):(5):[5]:(1)}; cuboctahedron {(3):6:[3]} = {(4):[4]:(4)}; anticuboctahedron {(2):(6):(2):[2]}.

L_{13}: edge-bridged icosahedron {(1):(2):[2]:(2):[2]:(2):[2]}; face-capped icosahedron {(1):(3):[3]:(3):[3]}; {(1):(5):(6):(1)}; {(3):[3]:(3):(4)}; {(1):(4): [4]:(4)}.

The more common ones are illustrated in Fig. 6.8.

As might be expected, the lowest energy geometries obtained by Benfield and Johnson were generally fully triangulated polyhedra since it is here that ligand-ligand repulsive forces are minimised. They then derived the structure of a cluster by simply placing one polyhedron (the metal) inside the other (the ligand). Often the proposed ligand envelope does not actually correspond to the observed ligand polyhedral form. The failures usually occur for systems in which the symmetry of the *idealised* metal polyhedron and that of the *idealised* ligand polyhedron are not immediately compatible. In such cases *each* polyhedron usually adjusts until a common symmetry is found. For example, the Co_6 unit in $[Co_6(CO)_{14}]^{4-}$ (Fig. 6.3) has O_h symmetry and the ligand polyhedron is an omnicapped cube, also of O_h symmetry, hence an O_h cluster is formed. In contrast, $[Co_6(CO)_{15}]^{2-}$ (Fig. 6.9) adopts the lower symmetry C_{3v} geometry where both the metal and ligand polyhedra are distorted from octahedral and triangulated respectively.

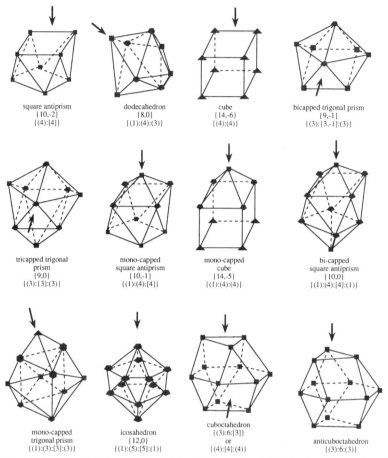

Fig. 6.8 Ligand polyhedra which minimise L-L repulsion, labelled using the notation of Benfield and Johnson (see text). The arrow indicates the "top" of the polyhedron.

Comparison with experiment showed that Benfield and Johnson's approach was a helpful way of proceeding, but not completely accurate. At about the same time Kepert and Williams developed the electron pair repulsion theory for determining ligand geometry.[39] This model is analogous to VSEPR and works well for close-packed ligand polyhedra, however, it has problems accounting for variations in geometry along series such as: $M_4(CO)_{12}$, M = Ir, Rh, Co. The Ir cluster has an anticuboctahedral ligand polyhedron (Fig. 6.8) whereas the other two are icosahedral with three bridging CO's. Kepert and Williams postulated either an increase in M-M bonding or different M-CO bonding to account for the Ir geometry. Johnson and Benfield [40] had previously noted that Ir_4 causes an expansion of the ligand polyhedron and concluded that the polyhedron rearranges to retain L-L contact - thus introducing an element of ligand-ligand attraction.

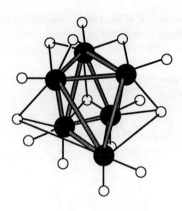

Fig. 6.9 $[Co_6(CO)_{15}]^{2-}$.

More recently Braga *et al.* have considered both repulsive and attractive interactions within ligand polyhedra. Using a model along the lines of the AAIM, they treated a series of carbonyls: $Mn_2(CO)_{10}$, $Fe_2(CO)_9$, $Co_2(CO)_8$ (Fig. 6.2), $Fe_3(CO)_{12}$, $Ru_3(CO)_{12}$, and two isomers of $Ir_6(CO)_{16}$ (Fig. 6.10). In general it was found that the C atoms in a given molecule are close packed and their interaction energy is repulsive, whereas the non-bonded O-C interactions are nearly always attractive and make a significant contribution to the stability of the molecules. A further conclusion was that the attractive component of the L-L interactions was an important contributor to the geometry adopted by the cluster, and also that crystal packing forces can have an important effect on the geometry observed in the crystal.

In conclusion, with this view of cluster bonding, we can say that the bonding in metal clusters can be understood as an interplay between trends towards (i) maximising the magnitude of the cohesive energy of the metal polyhedron, (ii) maximising M-L bonding energy, and (iii) avoiding steric crowding of the ligands. Any one of these factors can have a significant effect on the others, so that changing one can result in an entirely different cluster structure. For example, adding a ligand to form M_mL_{n+1} adds stability to the system due to the extra M-L bond (or bonds if the new ligand is a bridging one), destabilises the system by removal of d-electron density from M-M bonding in order to make the M-L bond(s), and alters the L-L interactions. Whether M_mL_{n+1} is more stable than M_nL_m is determined by the net effect of these energy changes. If it is less stable, then M_mL_{n+1} may decompose. Alternatively there may be a completely different geometry for which the total energy is lower. In this case, if there exists an allowed reaction pathway between the two structures whose activation energy is not excessive, then the system will spontaneously rearrange rather than decompose. If the activation energy is large compared with kT, then the rearrangement will not occur readily, but other methods of preparing M_mL_{n+1} may result in a more stable structure rather than the one which results from the addition of a ligand to M_mL_n. There are numerous examples of this type of behaviour to be found in the literature. One example is provided by the metal polyhedra of $Os_6(CO)_{18}^{2-}$, which is octahedral; if, however, $[HOs_6(CO)_{18}]^-$ is made, the metal polyhedron is a distorted octahedron, and $H_2Os_6(CO)_{18}$ has either a square based pyramidal or a distorted octahedral metal polyhedron.

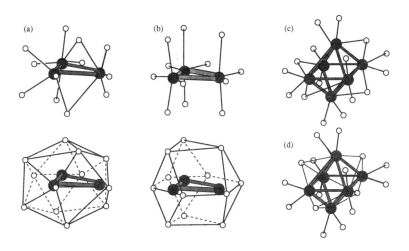

Fig. 6.10 (a) $Fe_3(CO)_{12}$, (b) $M_3(CO)_{12}$, M=Ru, Os, (c) black $Ir_6(CO)_{16}$, and (d) red $Ir_6(CO)_{16}$.

6.3 Some Examples

$M_2(CO)_n$

M_2 geometries, such as $Co_2(CO)_8$ and $[Fe_2(CO)_8]^{2-}$ illustrated in Fig. 6.2, are "precluster" structures, having only two metal atoms; however even such simple "clusters" serve to illustrate the virtue of maintaining both the ML_j fragment and M_m plus L_n polyhedral representations. Neither one is better, as both represent an aspect of the three dimensional reality. The cube of $[Fe_2(CO)_8]^{2-}$ is, according to Benfield and Johnson, more close packed than the bicapped trigonal prism, which reflects the smaller M-L bond lengths of this cluster relative to $Co_2(CO)_8$. We also see that packing effects over-ride the charge delocalization that bridging ligands normally provide.

$Fe_2(CO)_9$ (Fig. 6.2) has the same electron count as $[Fe_2(CO)_8]^{2-}$, but, it is neutral and the ligand polyhedron may be described as the *closo*-{9,0} structure. {9,0} is not completely defined by symmetry - it may be elongated or shortened along the three-fold axis unless an additional constraint such as uniform L-L distance is included. This molecule is elongated along the three-fold axis and the ligand connectivity in the polyhedron is more accurately described by linking the capping ligands rather than the ends of the prism, so becoming like two stacked octahedra sharing three vertices. $Mn_2(CO)_{10}$ adopts the *closo*-{10,0} geometry.

$M_3(CO)_n$ and fluxionality

The concept of a unique well-defined molecular geometry is appropriate for some clusters in all environments. However, some clusters are found to be constantly changing between different geometries (at least in a labelled-atom sense). We usually identify two extremes of this: (i) when the changes are too rapid for our experimental probes to identify distinct structures we call the behaviour fluxional, and (ii) when the different structures are long-lived we talk about isomerism. The demarcation between fluxionality and isomerism depends on the experimental technique being used. It is usually defined by NMR, though even here the line is not clear, and a useful rule of thumb is that rearrangements with an activation energy of 20 - 80 kJ mol^{-1} are fluxional at room temperature. In still other clusters we find that the geometry depends on environment, with quite different geometries being found in different phases or solvents. For more detailed discussions of fluxional behaviour of clusters see *e.g.* references [48,49].

$Os_3(CO)_{12}$ is an example of the rapidly interchanging behaviour. In solution below 60°C it has two peaks of equal intensity in the $^{13}\underline{C}O$ NMR spectrum as expected for the D_{3h} geometry illustrated in Fig. 6.10b.[41,42] However, above 60°C the $^{13}\underline{C}O$ NMR spectrum shows only one peak, though the geometry is still D_{3h}. $Os_3(CO)_{12}$ is thus fluxional (*i.e.* rearranges on the NMR timescale) at higher temperatures with all the ligands on average being found in the same environment. On the basis of Os-CO coupling, the probable explanation of this observation is that the equatorial and axial CO ligands on each Os undergo exchange, *i.e.* there is site exchange about the same Os at temperatures above 60°C, with the fluxionality being between two energetically degenerate geometries. In the solid a total of twelve independent chemical shift values are observed in accord with the lower symmetry imposed by the crystallographic lattice, but the structure still *approximates* to the same form independent of phase.

A good example of where the geometry depends on environment is $Fe_3(CO)_{12}$. In the solid state $Fe_3(CO)_{12}$ (Fig. 6.10a) has an icosahedral ligand polyhedron, but has been shown to exhibit two-fold disorder (*i.e.* the ligand and metal polyhedra randomly adopt either of two crystallographically equivalent relative orientations). This occurs because each molecule has carbonyl groups in asymmetric bridging positions along the shortest Fe-Fe edge (2.56 Å versus 2.68Å and 2.65Å); there are two posible (enantiomeric) arangements of these bridges and both exist in the unit cell.[43,44] Each molecule possesses approximate C_2 symmetry with the *pseudo* two-fold axis passing through the middle of the doubly bridged Fe-Fe bond and the opposite Fe atom. The disorder was initially explained in terms of a simple static model where some clusters had one geometry and others had the opposite handedness. However, in solution the structure appears to be solvent dependent. In non-polar solvents the infra red spectrum is consistent with nearly C_{2v} symmetry, so an average of the enantiomeric forms of the solid is found.[45] In more polar solvents there is a dramatic change. Bands associated with the CO-bridges decrease significantly in intensity and a different set of terminal bonds are apparent. This change is thought to be brought about by the conversion of the C_{2v} form into the D_3 form. This idea is appealing because the homologous $Ru_3(CO)_{12}$ and $Os_3(CO)_{12}$ have the related D_{3h} structure in the solid (Fig. 6.10b). Thus, it appears that $Fe_3(CO)_{12}$ exists in two structurally isomeric forms in solution, with the activation energy for the C_{2v} / D_3 rearrangement being less than 20 kJ mol^{-1} since ^{13}C NMR for enriched samples shows that all carbonyls are equivalent at -150°C. A possible mechanism for this was proposed to be *via* the libration of one polyhedron with respect to the other.[46] Hence, in this instance, it appears that the identity of the M-L bonds is not crucial, and a smooth very low energy transition from one type to another is possible. Since such a low energy transition is possible, it now appears that a dynamic explanation for the statistical disorder in the solid may be appropriate.

The rearrangement of cluster geometries is often described in terms of rapid ligand migration[47] or "scrambling". Although descriptive, such labels give an inaccurate impression since one gets a picture of randomly wandering ligands. The rearrangement processes always follow a well-defined vibration as discussed in §2.3.3. Often for clusters, what appears as a distinct breaking and forming of a large number of M-L bonds is in fact a simple relative rotation of the metal and ligand polyhedra with the gradual stretching of some and shortening of other M-L distances resulting in different M and L being described as bonded at the end of the process.

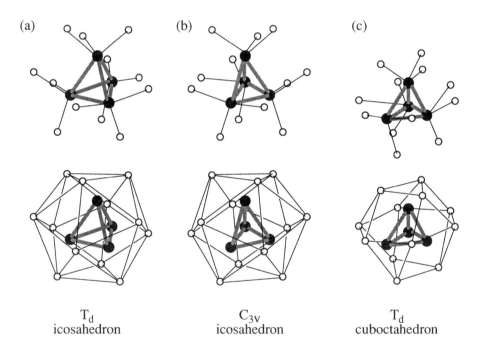

Fig. 6.11 Geometries of $M_4(CO)_{12}$.

$M_4(CO)_{12-n}L_n$

An example of the role of steric factors in determining molecular geometry comes from some Group 9 $M_4(CO)_{12-n}L_n$ clusters. These systems show isomerism of the ligand polyhedron in addition to isomers resulting from different relative orientations of the metal and ligand polyhedra. $Co_4(CO)_{12}$ has the smallest metal polyhedron and the ligands adopt the geometry expected if they are close-packed, namely {12,0}. The Co_4 tetrahedron has a number of possible orientations within the {12,0} ligand polyhedron, two of which are illustrated in Fig. 6.11. In the T_d

form (Fig. 6.11a) all the carbonyls are terminal. Rotation of the tetrahedron by 15° about any one of the four equivalent C_3 axes then generates the observed C_{3v} form (Fig. 6.11b). A libration between these two forms accounts for the fluxional behaviour observed in the solid state (see below). However, unlike the $Fe_3(CO)_{12}$ discussed above, there is no evidence for the T_d isomer (or any other) being a stable species, although low temperature infra red studies for crystalline samples indicate that some structural changes do occur.

$Rh_4(CO)_{12}$ also adopts the icosahedral ligand polyhedron with C_{3v} geometry for the whole cluster, as illustrated in Fig. 6.11b. However, when the Rh_4 is replaced by Ir_4, the ligand polyhedron is forced to expand and the cuboctahedron becomes a better arrangement for optimising L-L interactions (Fig. 6.11c). The metal tetrahedron is oriented so that the vertices of the tetrahedron coincide with alternate triangular faces of the cuboctahedron and all ligands are terminal. When one of the carbonyls of $Ir_4(CO)_{12}$ is replaced by $CN(t-Bu)$, little change in geometry is observed.[50] However, when four $P(CH_3)_3$ are substituted for carbonyls, the ligand polyhedron rearranges to form the C_{3v} edge-bridging icosahedral ligand polyhedron in order to accommodate the larger ligands.[51] The interchange between the cuboctahedron and icosahedron is shown in Fig. 6.12.

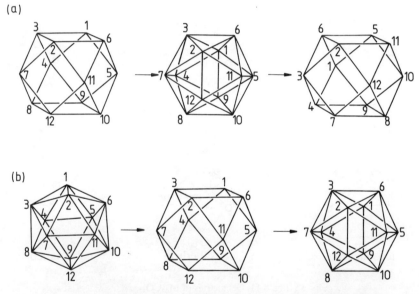

Fig. 6.12 Interchange between icosahedral and cuboctahedral ligand polyhedra.

Another driving force for structural change is the ligand geometry itself. If diars (Fig. 6.5) is substituted for two CO's, the requirements of the As-As bite size force the ligands to be chelated at a smaller angle than is consistent with the all terminal cuboctahedron. Thus in this case also the ligand polyhedron is icosahedral.[52] If $Me_2PCH_2CH_2PMe_2$ is substituted as the "chelating" ligand it bridges between two different Ir atoms,[53] but still forces an icosahedral ligand polyhedron because of its bulk.

M_6L_n

[Co$_6$(CO)$_{14}$]$^{4-}$ and Rh$_6$(CO)$_{16}$ are, as noted above, isoelectronic and have the same metal polyhedron geometry; but clearly they must have different ligand polyhedra (*cf.* Figs. 6.3 and 6.4). The former has six terminal and eight face-bridging carbonyls making a highly symmetric omnicapped cube ligand polyhedron {14,0}. The latter has twelve terminal and four face-bridging ligands in a {(1):(6):(3);[3]:(3)} polyhedron. However, there is more similarity between the geometry of these two different molecules than there appears to be at first sight. As noted by Benfield and Johnson, [Fe$_6$C(CO)$_{16}$]$^{2-}$ actually has the same metal polyhedron and same ligand polyhedron (the carbon is in the centre of the cluster) as Rh$_6$(CO)$_{16}$ (Fig. 6.4), however, their relative orientations make the molecule appear to be very different with the iron cluster having thirteen terminal and three asymmetric edge-bridged carbonyls. Presumably this allows the M-L bonds to be the correct length while maintaining L-L interactions, since edge-bridged ligands will be further from the metal core than face bridged-ones and so the former are more appropriate for the smaller iron metal polyhedron.

Finally, we consider two examples that are closely related, but for one of which no simplifying description of the geometry seems possible or even helpful. [Co$_6$C(CO)$_{13}$]$^{2-}$ (Fig. 6.13) [54] has thirteen ligands and unless thirteen ligands adopt the *closo* dodecahedron then usually no simple description is possible. However, if we take the liberty of coalescing the two shaded carbonyls in Fig. 6.13 into one bridging carbonyl, then the ligand polyhedron becomes cuboctahedron as is the case for Os$_3$(CO)$_{12}$ (though in that case there are six bridging ligands due to a different metal polyhedron). Encouraged by such a success one might expect to perform the same exercise on [Rh$_6$C(CO)$_{13}$]$^{2-}$ (Fig. 6.4).[55] However, due in large part to the steric requirements of the interstitial C and consequent distortions of the metal polyhedron (which is larger than for Co$_6$) and the need to distribute the ligand electrons over the cluster leads to a very irregular ligand polyhedron. Any description of the ligand polyhedron ends up being more complicated than the picture of the whole molecule, so in this instance it is pointless to pursue such a description of cluster geometry.

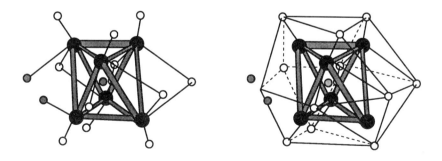

Fig. 6.13 [Co$_6$C(CO)$_{13}$]$^{2-}$. If the two shaded carbonyls are merged into one bridging ligand, then the ligand polyhedron may be described as a cuboctahedron.

References

(1) Belluco, U. *Organometallic and Coordination Chemistry of Platinum*; Academic Press: London, 1974.
(2) Cotton, A. F.; Wilkinson, G. *Advanced Inorganic Chemistry;* 5th ed.; Wiley-Interscience: New York, 1988.
(3) Greenwood, N. N.; Earnshaw, A. *Chemistry of the Elements;* 1st ed.; Pergamon Press: Oxford, 1984.
(4) *Comprehensive Organometallic Chemistry*; Wilkinson, G.; Stone, F. G. A.; Abel, E. W., Eds.; Pergamon Press: Oxford, 1982.
(5) Mingos, D. M. P.; Wales, D. J. *Introduction to Cluster Chemistry*; Prentice-Hall International, Inc.: New Jersey, 1990.
(6) Braga, D.; Rodger, A.; Johnson, B. F. G. *Inorganica Chimica Acts* **1990**, *174*, 185-191.
(7) Burdett, J. K. *Molecular Shapes: Theoretical Models of Inorganic Stereochemistry*; John Wiley and Sons: New York, 1980.
(8) Burdett, J. K. *Inorg. Chem.* **1978**, *17*, 47.
(9) Mason, R.; Thomas, K. M.; Mingos, D. M. P. *J. Amer. Chem. Soc.* **1973**, *95*, 3802.
(10) McPartlin, [.; Eady, C. R.; Johnson, B. F. G. *J. Chem. Soc. Chem. Commun.* **1976**, 833.
(11) Owen, S. M. *Polyhedron* **1988**, *7*, 253.
(12) Owen, S. M.; Brooker, A. T. *A Guide to Modern Inorganic Chemistry,* Longman: Singapore, 1991.
(13) Mingos, D. M. P.; Johnston, R. L. *Theoretical Models of Cluster Bonding*; Springer-Verlag: Germany, 1987; Vol. Structure and Bonding.
(14) Stone, A. J. *Inorg. Chem.* **1981**, *20*, 563.
(15) Stone, A. J.; Wales, D. J. *Inorg. Chem.* **1989**, *28*, 3120.
(16) Ceulemans, A.; Fowler, P. W. *Inorg. Chim. Acta* **1985**, *105*, 75.
(17) Mingos, D. M. P. *Nature Phys. Sci. (London)* **1972**, *236*, 99.
(18) Wade, K. *Chem. Comm.* **1971**, 792.
(19) Wade, K. *Adv. Inorg. Chem. Radiochem.* **1976**, *18*, 1.
(20) Mingos, D. M. P.; May, A. S. In *The Chemistry of Metal Cluster Complexes*; D. F. Shriver, H. D. Kaez and R. D. Adams, Eds.; VCH Publishers: New York, 1990.
(21) Evans, D. G.; Mingos, D. M. P. *Organometallics* **1983**, *2*, 435.
(22) Pettifor, D. G. In *Physical Metallurgy*; R. W. Cahn and P. Haasen, Eds.; Elsevier: Amsterdam, 1983.
(23) Woolley, R. G. In *Transition Metal Clusters*; B. F. G. Johnson, Ed.; John Wiley and Sons: Great Britain, 1980.
(24) Woolley, R. G. *Nouveau Journal de Chimie* **1987**, *5*, 219.
(25) Woolley, R. G. *Inorg. Chem.* **1985**, *24*, 3519.
(26) Johnson, B. F. G. *J. Chem. Soc. Chem. Commun.* **1986**, 29.
(27) Johnson, B. F. G.; Woolley, R. G. *J.C.S. Chem. Commun.* **1987**, 634.
(28) Sappa, E.; Tiripicchio, A. *Prog. Inorg. Chem.* **1987**, *35*, 437.
(29) Mingos, D. M. P. In *Comprehensive Organometallic Chemistry*; G. Wilkinson, F. G. A. Stone and E. W. Abel, Eds.; Pergamon Press: Oxford; Vol. 3.
(30) Elian, M.; Chen, M. M. L.; Mingos, D. M. P.; Hoffmann, R. *Inorg. Chem.* **1976**, *15*, 1148.
(31) Hoffmann, R. *Angew. Chem. Int. Ed.* **1982**, *21*, 711.
(32) Mingos, D. M. P. *J. Chem. Soc. Dalton* 133.
(33) Shaik, S.; Hoffman, R.; Fisel, C. R.; Summerville, R. H. *J. Amer. Chem. Soc.* **1980**, *102*, 4555.
(34) Woolley, R. G. *Inorg. Chem.* **1985**, *24*, 3525.
(35) Leung, P. C.; Coppens, P. *Acta Cryst.* **1983**, *B39*, 535.
(36) Albano, V. G.; Braga, D.; Martinengo, S. *J. Chem. Soc. Dalton* **1981**, 717.
(37) Johnson, B. F. G. *J.C.S. Chem. Comm.* **1976**, 211.
(38) Benfield, R. E.; Johnson, B. F. G. *Trans. Met. Chem.* **1981**, *6*, 131.
(39) Kepert, D. L.; Williams, S. C. *J. Organometallic Chemistry* **1981**, *217*, 235.
(40) Johnson, B. F. G.; Benfield, R. E. *J. Chem. Soc. Dalton Trans.* **1980**, 1743.

(41) Aime, S.; Gambino, O.; Milone, L.; Sappa, E.; Rosenberg, E. *Inorg. Chim. Acta* **1975**, *15*, 53.
(42) Forster, A.; Johnson, B. F. G.; Lewis, J.; Matheson, T. W.; Robinson, B. H.; Jackson, W. G. *J. Chem. Soc. Chem. Comm.* **1974**, *1042*.
(43) Wei, C. J.; Dahl, L. F.; Cotton, F. A.; Troup, J. M. *J. Amer. Chem. Soc.* **1969**, *91, 1351*.
(44) Wei, C. J.; Dahl, L. F.; Cotton, F. A.; Troup, J. M. *J. Amer. Chem. Soc.* **1974**, *96, 4155*.
(45) Johnson, B. F. G. *J. Chem. Soc. Chem. Commun.* **1976**, 703.
(46) Johnson, B. F. G.; Rodger, A. In *The Chemistry of Metal Cluster Complexes*; D. F. Shriver, H. D. Kaesz and R. D. Adams, Eds.; VCH Publishers inc: New York, 1990.
(47) Mingos, D. M. P.; Slee, T.; L.Zhenyang *Chem. Rev.* **1990**, *90*, 383.
(48) Bond, E.; Muetterties, E. L. *Chem. Rev.* **1978**, *78*, 639.
(49) Mann, B. E. In *Comprehensive Organometallic Chemistry*; G. Wilkinson, F. G. A. Stone and E. W. Abel, Eds.; Pergamon Press: Oxford, 1982; Vol. 3.
(50) Churchill, M. R.; Hutchinson, J. P. *Inorg. Chem.* **1979**, *18*, 2451.
(51) Darensbourg, D. J.; Baldwin-Zuschke, B. J. *Inorg. Chem.* **1981**, *20*, 3846.
(52) Churchill, M. R.; Hutchinson, J. P. *Inorg. Chem.* **1980**, *19*, 2765.
(53) Ros, R.; Scrivanti, A.; ALbano, V. G.; Braga, D. *J. Chem. Soc. Dalton Trans.* **1986**, 2411.
(54) Albano, V. G.; Braga, D.; Martinengo, S. *J. Chem. Soc. Dalton. Trans.* **1986**, 981.
(55) Albano, V. G.; Braga, D.; Martinengo, S. *J. Chem. Soc. Dalton. Trans.* **1981**, 717.

CHAPTER 7

Macromolecules: Small Changes and Large Effects

Contents

Introduction		163
7.1	DNA	164
7.2	Proteins	171
7.3	The final word	174

Introduction

The focus of the preceding chapters has been on the arrangement of two or more atoms about another atom (or, in the case of clusters, a polyhedron of other atoms) to which they are bonded. We have specifically avoided examining accurate molecular structures and have concentrated on the somewhat less precise phrase "molecular geometry", being concerned with the overall shape the molecules adopt. In this chapter we shall conclude our study of molecular geometry by looking briefly at two types of polymers - DNA's and proteins - where each subunit may be examined according to the previous chapters, but where small variations in subunit geometry lead to large changes in the overall molecular geometry and hence biological activity.

Many large books have been written about the geometry and structure of DNA's and proteins. Our aim here is to give merely a glimpse of the subject; the references at the end of the chapter will start a trail to a more detailed understanding. The structure of both types of molecule are usually discussed in terms of their primary, secondary and tertiary structures. *Primary structure* details which atoms are bonded to which atoms and is usually expressed as a list of connected nucleotides (DNA) or amino acids (protein); note that both types of molecule are made up of linear chains of a limited number of distinct subunits. Entropy, if nothing else, would ensure that proteins and nucleic acids did not remain as linear molecules in their natural aqueous environment. In fact energetic factors are also at work with the result that the molecules usually exist as well-defined structures. *Secondary structure* describes well-defined geometrical units of connected nucleotides or amino acids; this is akin to defining local structure in regions of the molecule, and its precise meaning will become clearer as we discuss examples in the following sections. *Tertiary structure* deals with the geometry of the molecule as a whole. Thus for DNA the primary structure is concerned with what phosphates, sugars, and bases are present; the formation of bases into base-pairs, and the angles between the bases is the province of secondary structure, while the tertiary structure relates to how the DNA helices are packed together. Our concern in previous chapters has been somewhere between primary and secondary structure. It should be stressed that the divisions between the different levels of structure is not clear and some blurring is inevitable when one tries to define them to precisely.

7.1 DNA

When asked the question "what is the geometry of DNA?", most people would either claim ignorance or talk about the double helical structure first proposed by James Watson and Francis Crick in 1953. The main features of the Watson-Crick double helical model of DNA are schematically illustrated in Figs. 7.1-3 and are as follows.

(i) DNA is composed of nitrogenous bases, deoxyribose sugar units, and phosphate groups. Each group of one base, one sugar and one phosphate is called a nucleotide. The base, either a purine (adenine and guanine) or pyrimidine (thymine and cytosine) derivative, is linked to the sugar *via* a glycosidic bond, and the phosphate to the sugar by a phosphoester bond (Fig. 7.1). There are four bases and hence four nucleotides in DNA. Nucleotides are then linked together by phosphoester bonds.

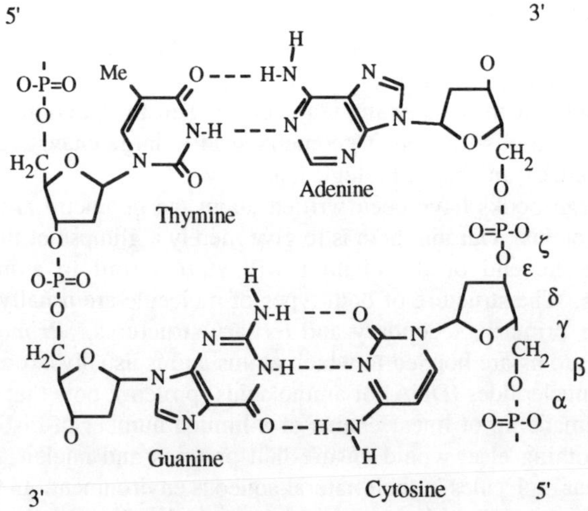

Fig. 7.1 Primary structure and common base-pairing for duplex DNA. The standard labels for the backbone torsion angles are indicated by Greek letters α - ζ along the bond about which the torsion occurs.

Macromolecules: Small Changes and Large Effects 165

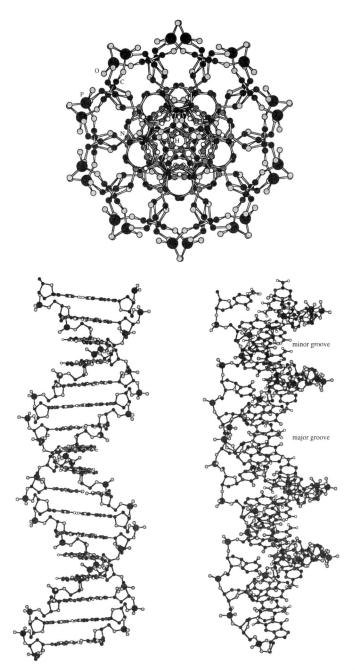

Fig. 7.2 Standard B-DNA from three perspectives, (a) looking down the helix axis and showing the width of the helix (this is twice the scale of the other perspectives), (b) perpendicular to the helix and showing the the right hand twist of the helix and base orientation relative to the helix axis, and (c) tilted to show the depths and widths of the major and minor grooves. Most of the hydrogens have been omitted for clarity.

(ii) The nucleotides are linked together to form polynucleotide chains. Two chains are held together by hydrogen bonds between two bases and are coiled round a common axis. Each base-pair of the nucleotide chains then forms a step in a ladder with each step linked by a sugar and a phosphate. Successive nucleotides have the 5'-hydroxyl of one sugar linked *via* the phosphate and a phosphodiester bond to the 3'-hydroxyl of the next sugar. Thus each strand of the DNA has two distinct ends, one having a 5'-hydroxy group not linked to a nucleotide and the other an unlinked 3'-hydroxy group. By convention the base sequence is read from 5'- to 3'-, although this convention is not always heeded.

(iii) The chains of the double helix run in opposite directions and form a right-handed helix. The bases are located on the inside of a chain of the phosphate and deoxyribose units that form the backbone (see large black circles in Figs. 7.2). The planes of the bases are perpendicular to the helical axis and may be identified in Fig. 7.2 from the position of the nitrogens.

(iv) Adenine (A) is always paired with thymine (T), and cytosine (C) with guanine (G). The AT base-pairs are stabilised by two hydrogen bonds, and GC pairs by three hydrogen-bonds (Fig. 7.1). The π-π interactions between stacked bases on the same strand also stabilises the double helix.

(v) The helix is 2.0nm in diameter. Adjacent bases are separated by 0.34nm along the axis and related by a rotation of 36°. The helical repeat is therefore 10 residues or 3.4nm.

(vi) Because the glycosidic bonds of a base pair (*i.e.* the one that joins it to the sugars) are not symmetric about the helical axis, two kinds of groove are present - the major groove (width 1.17nm and depth 0.85nm) and the minor groove (width 0.57nm and depth 0.75nm).

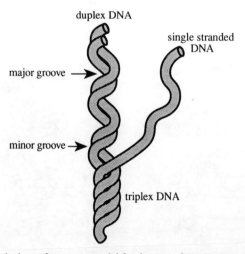

Fig. 7.3 A piece of rope as a model for the secondary structure of DNA.

The Watson-Crick description of the geometry of DNA contains both the primary structure and the secondary structure. It is most easily visualised by taking a piece of rope, which is typically composed of three strands twisted into a right-handed helix, and removing one of the strands. The resulting right-handed double helix has a major groove (where the strand of rope has been removed) and a minor groove (Fig. 7.3). The Watson-Crick model had such a revolutionary effect on biology because it is such a simple model that is very close to being correct. DNA is generally found in the double helical, or *duplex*, form. Further, the most common duplex DNA structure is B-DNA (Fig. 7.2) which has bases more-or-less perpendicular to the helical axis and the backbone following a right-handed twist.

Table 7.1 Typical geometric parameters for standard A-, B-, and Z-forms of DNA.[3-7] The repeat unit of Z-DNA is a dinucleotide, necessitating two values of each angular parameter. ζ adopts a range of values in Z-DNA.

	A-DNA	B-DNA	Z-DNA
α	-85°	-47°	60° / 160°
β	-152°	-146°	-175° / -135°
γ	46°	36°	178° / 57°
δ	83°	156°	140° / 95°
ε	178°	155°	-95° / -110°
ζ	-46°	-95°	-35° - 85°
sugar conformation	$C_{3'}$-endo	$C_{2'}$-endo	$C_{3'}$-endo / $C_{2'}$-endo
glycosidic bond	*anti*	*anti*	*anti* (C), *syn* (G)
base roll	12°	0°	1°
base tilt	20°	5°	9°
base twist	32°	36°	11° / 50°
base slide	0.15nm	0nm	0.2nm
helix diameter	2.55nm	2.37nm	1.84nm
bases / turn of helix	11	10	12
base rise/base pair	0.23nm	0.33nm	0.38nm
major groove	narrow, deep	wide, deep	flat
minor groove	broad, shallow	narrow, deep	narrow, deep

Although essentially correct, this model has proved to be rather too simplistic. Crystal structures of dodecanucleotides have shown significant local deviations from the Watson-Crick model with rotation angles varying from 28° to 42° and groove widths being slighty dependent on the base sequence. In addition, DNA is a polymorphic molecule and several classes of right-handed (A, B, C and D) and left-handed (Z) double helices have been identified. Which one (or ones, as there is growing evidence that in solution an equilibrium exists between a number of polymorphs of DNA) is adopted depends on the conditions and, to a lesser extent, on the sequence of the DNA. The B-form (Fig. 7.2) is the most common in solution form. It is a right-handed helix with base-pairs stacked approximately perpendicular to the long helical axis of the DNA. However, it should be noted that recent linear dichroism experiments by Johnson and coworkers[1] have shown B-form DNA to have bases tilted about 20° from being perpendicular to the helix axis. Further, successive bases can also slide sideways with respect to one another or roll or twist as illustrated in Fig. 7.4. Under high salt or low water (in solution usually this means high ethanol) conditions, B-DNA changes into the A-form (Fig. 7.5) which is a right-handed helix with the bases approximately parallel to each other but tilted at an angle of 70° to the helix axis. A-DNA has 11 base-pairs per turn of the helix and is wider than B-DNA, which has 10 - 10.5 base-pairs. These features are shown in Figs. 7.2 and 7.5. Further, the minor groove is essentially absent from A-DNA as shown in Fig. 7.5. Some helix parameters for different DNA polymorphs are collected in Table 7.1.

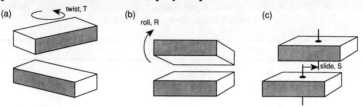

Fig. 7.4 (a) Twist, (b) roll, and (c) slide of DNA base-pairs with respect to one another.

It had initially been thought that DNA was intrinsically right-handed due to the local geometry of the sugars (the only chiral subunit of DNA). So it was with some surprise in 1979 that a crystal structure of Z-DNA was solved (Fig. 7.6).[2] This structure has DNA as a left-handed helix with the bases approximately perpendicular to the helix axis, but unlike B-DNA it does not have the same regular twist between consecutive base-pairs. Instead two base-pairs are stacked almost vertically, and then the next two are noticably zig (or zag) to the first pair. Twelve residues are found per turn of the helix. Z-DNA has the base-pairs shifted into the major groove, so that it is really no longer a groove.[3] Z-DNA is more likely to occur with G-C base-pairs, though it is not sequence specific as was intially thought. For example, poly[d(A-T)]$_2$, which is a double helical DNA with alternating A and T along each strand, adopts a left-handed form in the presence of Ni^{2+}.

The intriguing thing about the variations in Table 7.1 between A- and B-DNA are that they can be seen to arise from the small difference in sugar puckers. Four vertices of the pyranose ring are planar and the fifth is only about 0.05 nm from the plane. Which C is out-of-plane differs between the two polymorphs, with Z-DNA resulting from the *syn* orientation of half of its glycosidic bonds (Fig. 7.7). Since biological function requires the interaction of DNA with other molecules, in particular with proteins which have their own particular shapes as discussed below, both the large and small variations that occur in the molecular geometry of DNA may be crucial for its biological activity. The larger differences are from torsion angles adopting, say, *gauche* instead of *trans* forms. As discussed in §3.2.1 the energy difference between two conformers is small unless there is a steric clash. So despite the above discussion of the definitions of DNA geometry, its most important features probably relate to its flexibility and non-rigidity. Even the hydrogen-bonding between the bases that stabilises the duplex structure show many small but significant variations beyond the standard adenine-thymine and guanine-cytosine links.

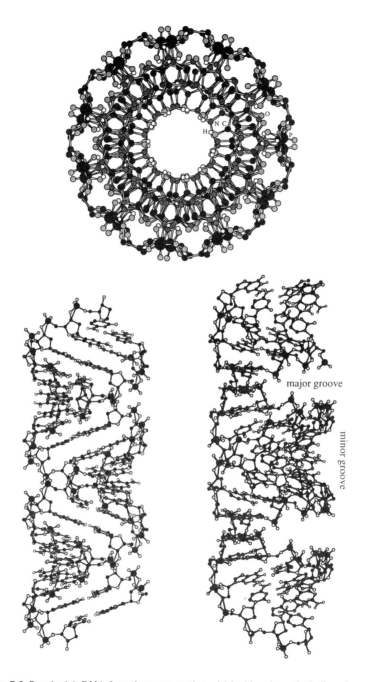

Fig. 7.5 Standard A-DNA from three perspectives, (a) looking down the helix axis and showing the width of the helix (this is twice the scale of the other perspectives), (b) perpendicular to the helix and showing the right hand twist of the helix and base orientation relative to the helix axis, and (c) tilted to show the depths and widths of the major and minor grooves. Most of the hydrogens have been omitted for clarity.

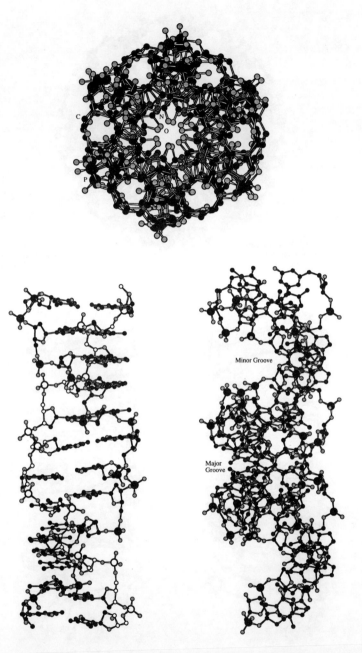

Fig. 7.6 Standard Z-DNA from three perspectives, (a) looking down the helix axis and showing the width of the helix (this is twice the scale of the other perspectives), (b) perpendicular to the helix and showing the left hand twist of the helix (the P's and C's of one backbone have hollow atoms) and base orientation relative to the helix axis, and (c) tilted to show the depths and widths of the major and minor grooves. Geometries were kindly supplied by Dr. I.S. Haworth. Most of the hydrogens have been omitted for clarity.

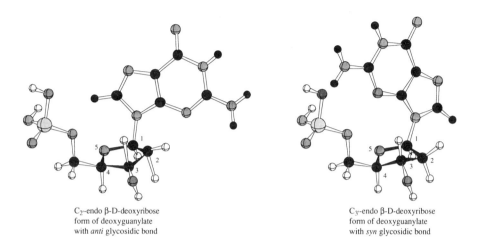

C$_{2'}$-endo β-D-deoxyribose
form of deoxyguanylate
with *anti* glycosidic bond

C$_{3'}$-endo β-D-deoxyribose
form of deoxyguanylate
with *syn* glycosidic bond

Fig. 7.7 C$_{3'}$-endo and C$_{2'}$-endo 2-deoxyribose, with *anti* and *syn* glycosidic bonds. *Anti* refers to the fact that the sugar and the base are on opposite sides of the bond.

Within a typical human cell, which is only 10^{-5}m in diameter, there can be as much as 2m in length of DNA. Obviously, the DNA must be packed very tightly, and this is achieved in several stages. First, the DNA is wrapped around proteins called histones; to continue the rope analogy raised earlier (*cf.* Fig. 7.3), this is like the way rope is wrapped around the capstan in a ship. These "capstan" units are then linked together by linear strands of DNA to resemble a beaded necklace, and finally the "necklace" itself is wrapped up even further. After so much compacting of the DNA it is amazing to realise that it still manages to unfold and then re-fold whenever it is required for cell processes. A feature of DNA that facilitates these processes is its ability to supercoil. Supercoiling can best be explained by a simple experiment. Take a 10cm piece of knitting wool (which is usually a left handed triplex) and untwist it. Then join the ends together and let it retwist itself; the wool will end up recoiling to less than its original length. The original helical twist is reinstated together with negative supercoiling. Positive supercoiling may be achieved by twisting the original pieces of wool further in the direction of its natural twist before joining the ends together.

7.2 PROTEINS

Although the functions and geometries of proteins and nucleic acids (both DNA and RNA) are very different, these molecules are built on the same principles. Both are large and complex yet are composed of a linear chain of smaller repeating units. In both cases, one of the main driving forces towards secondary structure formation is hydrogen-bonding.

The repeat unit of a protein is an amino acid, whose general formula is as shown in Fig. 7.8. There are twenty different amino acids to be found in proteins, they differ only in the identity of their R group (see Fig. 7.8). Naturally occurring amino acids all have the same handedness, as illustrated, and are referred to as L-amino acids. They are joined together by "*peptide bonds*" formed between the acid functional group of one amino acid and the amine group of another (Fig. 7.8) with the elimination of a water molecule. One end of the protein molecule is thus the amino end (by convention this is the beginning) and the other end is the carboxylate end; these are usually referred to as the N and C termini respectively. The molecular identity of the protein is given by listing the sequence of amino acids.

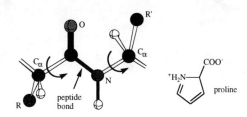

amino acid	R	amino acid	R	amino acid	R
glycine	-H	isoleucine	—C—C—C(CH₃)(CH₃) / H H CH₃	glutamic acid	—C—C—C(OH)(=O) / H₂ H₂
alanine	-CH₃				
valine	—C(H)(CH₃)(CH₃)	asparagine	—C—C(NH₂)(=O) / H₂	glutamine	—C—C—C(NH₂)(=O) / H₂ H₂
cysteine	—C(H₂)—SH	phenyl-alanine	—C(H₂)—C₆H₅	lysine	—(CH₂)₃—N(H)—C(=NH₂⁺)—NH₂
serine	—C(H₂)—OH	aspartic acid	—C(H₂)—C(OH)(=O)	tyrosine	—C(H₂)—C₆H₄—OH
threonine	—C(H)(OH)(CH₃)	leucine	—C(H₂)—C(H)(CH₃)(CH₃)	methionine	—C—C—S—CH₃ / H₂ H₂
histidine	imidazole-CH₂	tryptophan	indole-CH₂	arginine	—(CH₂)₃—N(H)—C(=NH₂⁺)—NH₂

Fig. 7.8 Amino acids and the peptide bond.

Proteins form regular secondary structural units essentially because the peptide unit O=C-N- is planar and rigid, but there is a large degree of rotational freedom about its links to the rest of the protein chain. Consequently, the polypeptide chain can arrange itself so that the C-O of one peptide unit hydrogen-bonds with an N-H of another unit in a number of different ways. The common secondary structure features are the right-handed α-helix (Fig. 7.9) and the β-sheet (both parallel and anti-parallel versions, Fig. 7.10). Many other structural units may also be identified, depending upon the detail with which crystal structures are examined. The α-helix is a rod-like structure where the nth peptide hydrogen-bonds its C-O to the N-H of the $(n+4)$th peptide and its N-H to the $(n-4)$th CO (Fig. 7.9). It thus forms a right-handed helix with 1.5Å translation and 100° turn between two consecutive peptides, giving 3.6 amino acid residues per turn. The alternative efficient formation of hydrogen bonds occurs between a sheet of parallel or antiparallel runs of amino acids; this is known as a β-sheet (Fig. 7.10). Typically the strands of an anti-parallel β-sheet are linked by β-turns where the nth peptide hydrogen-bonds with the $(n+3)$rd peptide. If a β-sheet extends over more than two strands, then the relative arrangements of the strands in space must be considered. They are often twisted with respect to one another.

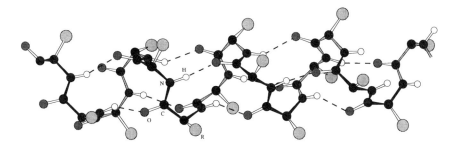

Fig. 7.9 α-helix structure. Dotted lines indicate hydrogen bonds.

In biological systems the α-helices, β-sheets, β-turns and other structural forms are packed together in very stable conformation. The way this is achieved generates two types of protein: globular and membrane. In the former, one of the driving forces for adopting the tertiary structure is to ensure hydrophobic side chains are wrapped up by the protein while hydrophilic side chains interact with water.[†] In membrane proteins, the converse is usually true as the protein is "solvated" not by water but by the hydrophobic membrane. Other factors that determine tertiary structure include the stabilising effect of hydrogen bonds and disulfide S-S bridges between amino acids that are remote from one another, usually in different secondary structural units.

One of the difficulties in comprehending the geometry of a protein is that they are so large that the overall geometry can be lost in the details. So, even more than in the case of clusters, we need a means of representing proteins that conveys the maximum amount of geometric information with minimal confusion. Protein chemists have adopted various illustration conventions, the most common of which is the ribbon diagram in which α-helices are represented by helical twirls of ribbon, β-sheets by wider flat pieces of ribbon with arrows on the C terminus, and often disulfide bridges are indicated by a zig-zag (Fig. 7.11). Although a lot of detail is lost, especially in the β-sheets since which residues are connected by hydrogen bonds is not indicated, such a diagram does still carry a

---

† Proteins are usually found in an aqueous environment.

great deal of information in a form that is possible to digest visually. More detailed pictures may then be tackled if required.

A protein's secondary and tertiary structure is crucial to its biological activity. In nature proteins appear to adopt unique folded conformations that optimise this activity. However, the probability of a typical 100-200 amino acid protein folding spontaneously to the correct structure by chance is essentially zero, and so an active area of research is to discover and be able to control how proteins fold. Only when the process of protein-folding is understood will we be able to contemplate designing proteins for specific biological applications.

parallel β-sheet

anti-parallel β-sheet

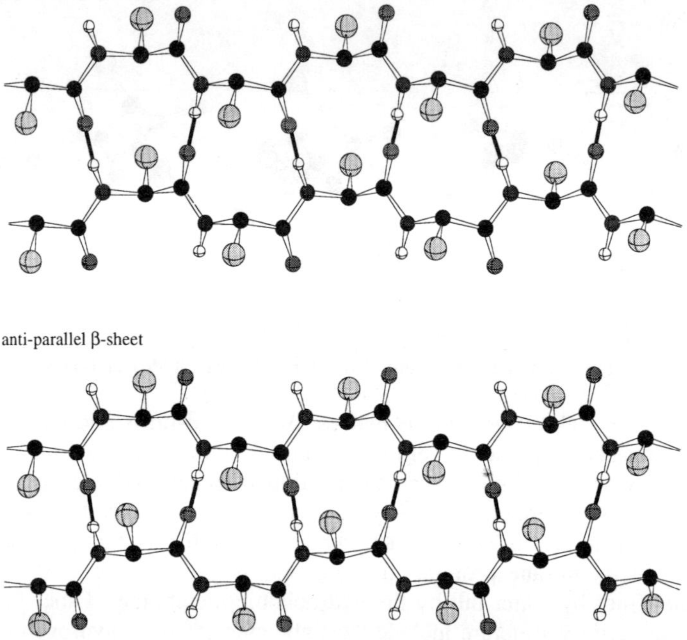

Fig. 7.10 Parallel and anti-parallel β-sheets. Hydrogen bonds between peptide chains are indicated by the thick lines. Atom / group labelling as for Fig. 7.9.

7.3 The Final Word

One of the themes running through the previous chapters has been the interplay of steric and electronic factors and how we might discern when one or the other is dominant. In determining the secondary and tertiary structure of macromolecules, it seems that electronic factors may be almost completely ignored. Significant success has been had with modelling the geometry of DNA and protein molecules using molecular mechanics (§1.3.2) which includes electronic factors only in a local sense: they determine preferred bond lengths, bond angles, and torsion angles, and specify how easily these can be deformed.

Hydrogen-bonding is modelled as a simple electrostatic effect. It currently seems that the limitations and failures of this approach are due to inadequacies of the parameters chosen rather than the fundamentals of the molecular mechanics method itself. Interactions of macromolecules with other molecules also seem to be amenable to molecular mechanics and related techniques. The word "seems" in both preceeding sentences should be noted, especially given the discussion of stereoelectronic effects on sugars in Chapter 3. It is, however, fortunate that steric effects account for so much of the macromolecular secondary and tertiary stucture as it encourages us to continue pursuing molecular modelling techniques to aid in our understanding of the structures of biological molecules, the design of DNA and protein binding drugs, and in the development of enzyme based chemistry.

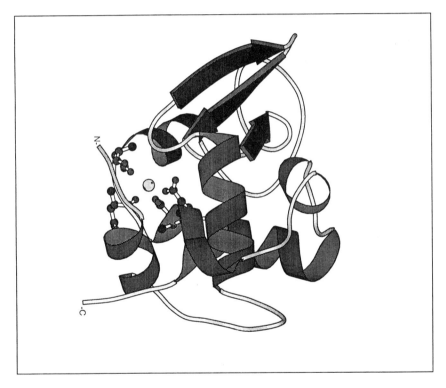

Fig. 7.11 Ribbon drawing of baboon α-lactalbumin with bound Ca showing the four ligating aspartates explicitly. Adapted from a diagram kindly provided by S. E. Radford.

References

(1) Chou, P.-J.; W.C. Johnson, J. *J. Amer. Chem. Soc* **1993**, *115*, 1205.
(2) Wang, A. H.; Quigley, J.; Kolpak, G. J.; Crawford, J. L. *Nature* **1979**, *282*, 680.
(3) Egli, M.; Williams, L. D.; Gao, Q.; Rich, A. *Biochemistry* **1991**, *30*, 11388.
(4) Stryer, L. *Biochemistry;* 3rd ed.; W.H. Freeman and Company: New York, 1988.

(5) Calladine, C. R.; Drew, H. R. *Understanding DNA, The molecule and how it works*; Academic Press Ltd: Cambridge, 1992.
(6) Beveridge, D. L.; Jørgensen, W. L. *Annals of the New York Academy of Sciences* **1986**, *482*.
(7) Gessner, R. V.; Frederick, C. A.; Quigley, G. J.; Rich, A.; Wang, A. H.-J. *J. Biol. Chem.* **1989**, *264*, 7921.

APPENDIX 1

Rules for Multiplication of Point Symmetry Operations

Introduction

The product of two symmetry operations, $(R_1 \times R_2) = (R_1R_2)$, is defined to be the operation of R_2 on the molecule or other object, followed by operation of R_1. Note that we define multiplication to proceed from right to left. The results of multiplying two symmetry operations together may be summarised by the following eight rules given below. The rules form the basis for the generating of point groups as outlined in §2.1.2 and discussed further in Appendix 2. The first five rules are apparent upon inspection. Rule 8 may be determined from reference[1] and Rules 6 and 7 may be proved by expressing the rotations as matrices operating on Cartesian vectors, and evaluating the products explicitly. We use the notation $\text{lcm}[m,n]$ to denote lowest common multiple of m and n (so for example $\text{lcm}[3,4] = 12$, but $\text{lcm}[3,6]=6$) and superscripts to denote the axis about which a rotation operates, or which is the normal (perpendicular vector) to a reflection plane. If the axis is unspecified then the default is the z axis, which is usually the major rotation axis. The notation for the symmetry operations is defined in §2.1.1.

Rule 1: A group with elements C_m^α and C_n^α must also contain C_r^α, $r = \text{lcm}[m,n]$. For example, if C_3^z and C_4^z are elements of a group then C_{12}^z is also a symmetry element as $(C_4)^{-1}C_3^z = C_{12}^z$. In general, this follows is because if C_m^α is in a group, then so is $(C_m^\alpha)^{-1}$, and $(C_m^\alpha)^{-1}(C_n^\alpha) = C_r^\alpha$

Rule 2: A group with elements C_m^α and S_n^α must also contain S_r^α, $r = \text{lcm}[m,n]$

Rule 3: A group with elements C_2^β and C_2^γ, where β and γ are separated by an angle θ and are both perpendicular to an axis α, must also contain C_r^α, $r = \pi q/\theta$, where q is the smallest integer such that $\pi q/\theta$ is an integer.

A corollary of this is that a group with elements C_2^β and C_r^α must also contain r lots of C_2 axes all perpendicular to α, at angles π/r from their nearest C_2 axis.

Rule 4: A group with elements σ^β and σ^γ must also contain C_r^α, $r = \pi q/\theta$; β, γ, and q as in Rule 3. Conversely, a group with elements σ^β and C_r^α must also contain r reflection planes with axes perpendicular to α (so the plane contains α) and at angles π/r from their nearest plane.

Rule 5: A group with elements C_2^β and σ^γ must also contain S_r^α, $r = \pi q/(\pi/2-\theta)$, where all labels are as in Rules 3 and 4. The converse of this depends upon whether r is an even or an odd number, and is discussed below.

Rule 6: R_ξ^α and R_η^β commute, *i.e.* $R_\xi^\alpha \times R_\eta^\beta = R_\eta^\beta \times R_\xi^\alpha$, if and only if one of the following conditions hold:
(i) α and β are either parallel or antiparallel,
(ii) α and β are perpendicular; and $(R_\xi^\alpha)^2 = (R_\eta^\beta)^2$.

Rule 7: R_ξ^α and R_η^β, where the angle between α and β is θ, satisfy $(R_\xi^\alpha \times R_\eta^\beta)^2 = E$, if and only if one of the following conditions hold
(i) $(R_\xi^\alpha)^2 = (R_\eta^\beta)^2 = E$ and $\theta = 0$, $\pi/2$, or π (*i.e.* α and β are parallel, antiparallel, or perpendicular),
(ii) Either $(R_\xi^\alpha)^2 = E$ or $(R_\eta^\beta)^2 = E$, $\theta = \pi$, and $|\xi-\eta| = \pi \pmod{2\pi}$, where ξ is the angle through which R_ξ^α rotates the system about the axis α and similarly R_η^β,
(iii) $(R_\xi^\alpha)^2 \neq E$, $(R_\eta^\beta)^2 \neq E$, for R_ξ^α and R_η^β both proper operations, and $\cos\theta = \cot(\xi/2)\cot(\eta/2)$,
(iv) $(R_\xi^\alpha)^2 \neq E$, $(R_\eta^\beta)^2 \neq E$, for R_ξ^α and R_η^β both improper operations, and $\cos\theta = \tan(\xi/2)\tan(\eta/2)$,
(v) $(R_\xi^\alpha)^2 \neq E$, $(R_\eta^\beta)^2 \neq E$, for R_ξ^α a proper operations, R_η^β an improper operation, and $\cos\theta = -\cot(\xi/2)\tan(\eta/2)$.

Rule 8: The group generated by R_ξ^α and R_η^β with its multiplication table constrained by
$$(R_\xi^\alpha \times R_\eta^\beta)^k = (R_\xi^\alpha)^m = (R_\eta^\beta)^n = E$$
is finite if and only if
$$1/k+1/m+1/n > 1.$$

Reference

(1) Coxeter, H. S. M.; Moser, W. O. J. *Generators and Relations for Discrete Groups*; Springer-Verlag: 1957.

APPENDIX 2

Generating Point Groups

Introduction

There are many ways we could generate the point groups using an augmentation procedure and starting from the groups that can be generated by one operation. The one we shall follow is designed to show the relationships between different point symmetry groups and hence should indicate how results for one point group need to be modified for other related groups. We begin with the groups generated by one proper operation, then augment these by the addition of another proper operation to generate all the chiral (*i.e.* asymmetric, §2.2) point groups. The achiral point groups then follow by addition of an improper operation. The multiplications rules of Appendix 1 are referred to by number.

A2.1 Chiral Point Groups

C_n: The set of all multiples of C_n (*i.e.* C_n carried out iteratively) is the cyclic group of order n, denoted C_n. The operation $(C_n)^k$ where n and k are coprime also generates the same group. By Rule 1, any combination of proper rotations about the z or $-z$ axes generates one of these groups.

D_n: If $\{C_n\}$ is augmented by C_2^x, where x is an axis perpendicular to z, then the D_n groups are generated. D_n contains an n-fold rotation about z, and n C_2^β rotations about axes perpendicular to z with $\beta = [\cos(2\pi k/n)]x + [\sin(2\pi k/n)]y$, $k = 0, ..., (n-1)$ by Rule 3.

If z is an infinitesimal rotation axis, then C_∞ and D_∞ are generated. We now have all the groups that satisfy Rules 6, or 7(i), or 7(ii) with respect to all other operations in the group.

The pairs of operations that satisfy Rule 7 (iii) and Rule 8 generate the remaining finite chiral groups. There are only three different ones of these.

T: If $m = n = 3$ in Rule 8 with the angle between the two C_3 axes being $\cos^{-1}(1/3)$, then the product of the two operations is a two-fold rotation axis and the tetrahedral group **T** is generated. The best way to visualise the tetrahedral group is to inscribe four points into alternate corners of a cube and to ignore the improper rotations. Alternatively, to generate **T** itself, replace each point by a tripod with bent feet. The three-fold axes are then along the vectors $(\pm 1, \pm 1, \pm 1)$, and the two-fold axes are x, y and z.

O: If $m = 3$ and $n = 4$ in Rule 8 with the angle between the C_3 and C_4 axes being $\cos^{-1}(1/\sqrt{3})$, then the octahedral group **O** is generated. The basic symmetry of this shape is best visualised by putting vertices in the middle of the faces of a cube an ignoring the improper rotations. Alternatively, replace the points by bent crosses to generate **O** itself.

I: If $m = 3$ and $n = 5$ in rule 8 with the angle between the C_3 and C_4 axes being $\cos^{-1}[1/\sqrt{3}\cot(\pi/5)]$, then the chiral icosahedral group, **I**, is generated. This group actually contains **T** as a subgroup, as can be seen by drawing a tetrahedron inside an icosahedron (*cf.* $M_4(CO)_{12}$ in Fig. 6.11).

Σ: Any pair of operations that satisfies only Rule 7(iii) generates the inifinite chiral spherical group, Σ.

A2.2 Achiral Point Groups

Achiral point groups are most simply generated by augmenting the chiral groups with improper operations that generate no new proper operations. This is always possible since the proper operations of an achiral point group form one it its subgroups.

Augmentation by the inversion i
C_{2nh}, $S_{(2n+1)}$, D_{2nh}, $D_{(2n+1)d}$, T_h, O_h, I_h, Σ_h: From Rule 6 we know that i commutes with all other point operations. Addition of i to a chiral group therefore adds no new proper operations and doubles the size or order of the group. The groups listed above result from the augmentation of the point groups C_{2n}, $C_{(2n+1)}$, D_{2n}, $D_{(2n+1)}$, **T**, **O**, **I**, Σ, respectively, by the operation i. The label C_{2nh} means the group contains C_{2n} and a σ_h; $S_{(2n+1)}$ contains all the multiples of the operation $S_{(2n+1)}$; D_{2nh} contains, among other operations, those of D_{2n} and $2n$ σ_v planes containing the C_2 axes; $D_{(2n+1)d}$ is also generated from D_m; however, it is quite different from D_{2nh} as it has its vertical reflection planes, σ_d, *bisecting* the C_2 axes

Generating Point Groups 181

and also has $S_{2(2n+1)}$; T_h, O_h, I_h, are Σ_h simply the achiral versions of their chiral counterparts.

Augmentation by a reflection plane σ

In order for σ to generate no new proper operations it must transform proper rotation axes into the same or other axes of the same order. Thus, any augmenting plane that must generates no new proper operations must either contain any axis, or be perpendicular to it, or bisect two rotation axes of the same order.

C_{nh}, D_{nh}: Addition of a horizontal reflection plane whose axis is parallel to the C_n axis of C_n and D_n generates, respectively, C_{nh} and D_{nh}. D_{nh} also has σ_v planes, and may be generated by augmenting D_n with a reflection plane that contains both C_n and one of the C_2 axes of D_n.

D_{nd}: Augmentation of D_n by a dihedral reflection plane, σ_d, that has its axis perpendicular to C_n (so the plane contains C_n) and bisects two neighbouring C_2 axes of D_n leads to D_{nd}. As noted above, D_{nd} contains S_{2n}, so D_{nd} is a higher order rotation group than its label indicates, but the maximum order rotation is improper not proper.

T_d: Two types of reflection plane may be added to T to generate achiral groups that do not have any additional proper operations. Addition of a plane whose axis is parallel to one C_2, contains the other two C_2 axes, and exchanges (and inverts) two pairs of C_3 axes leads to T_h. Alternatively, a reflection plane containing two C_3 axes and one C_2 axis also generates no new proper operations. The resulting point group in this case is T_d, of which methane is a common example.

Augmentation by an improper rotation

The only new point groups that augmentation by S_n, $n > 2$, can yield are ones without i or σ. So by Rules 2 - 5 we conclude that the final point group we require is S_{4n} resulting from the augmentation of C_{2n} by S_{4n}. It sometimes also is convenient to view D_{2nd} as resulting from the augmentation of D_{2n} by S_{4n}, and T_d as resulting from the augmentation of T by S_4.

APPENDIX 3

The Jahn-Teller Theorem

The proof of the Jahn-Teller theorem is seldom reproduced, but it is one of the simplest symmetry proofs, especially in the original form as produced by Jahn and Teller. The proof also shows why the theorem holds and what its consequences are.

Let H be the electronic Hamiltonian of the system. At any point, Q_o, on the potential energy surface, a Taylor Series expansion with respect to any normal coordinate Q^v, may be written:

$$H(Q^v) = H(Q_o + \delta Q^v) = H(Q_o) + \frac{\partial H}{\partial Q^v}\Big|_o \delta Q^v + \ldots$$

Using perturbation theory and staying within the Born-Oppenheimer approximation, the potential energy is

$$E(Q_o + \delta Q^v) = E(Q_o) + \left\langle \psi \left| \frac{\partial H}{\partial Q^v}\Big|_o \right| \psi \right\rangle_o \delta Q^v + \ldots$$

where ψ is the molecular wavefunction for the system at Q_o. If Q_o is a minimum energy point, then $\frac{\partial E}{\partial Q^v}\Big|_o = 0$ for all coordinates Q^v, which means that either $\frac{\partial H}{\partial Q^v}\Big|_o = 0$ or the spatial symmetry of the integral is such that it cancels out and gives a value of zero, *i.e.*

$$A_{1g} \not\subset \Gamma_\psi \times \Gamma_{\partial H/\partial Q^v|_o} \times \Gamma_\psi = \Gamma_\psi \times \Gamma_{Q^v} \times \Gamma_\psi$$

What Jahn and Teller showed by determining the symmetries of all vibrations for all molecules (their paper is a very good general reference for this reason) was that for non-linear molecules at least one non-totally symmetric vibration has the symmetry such that

$$\Gamma_\psi \times \Gamma_{Q^v} \times \Gamma_\psi = A_{1g}$$

for all degenerate ψ. Thus, if the electronic state of a non-linear molecule is degenerate, then at least one molecular vibration takes the molecule to a lower symmetry but more stable geometry.

INDEX

α-helix, *see* protein
3-acetoxy-tetrahydropyran, 80
AAIM: *see* atom-atom interaction model
achiral, 39, 180
actinides, 132*f*
activation volume, 63
adamantanones, 83
adenine, 164*f*
Al: *see* Group 13
alums, 131
amino acid, 6, 163, 172*f*
anomer, 81
anomeric effect, 81
anti-bonding: *see* orbital
antiperiplanar, 83
antisymmetrised product, 57
ao: *see* atomic orbital
arachno: *see* polyhedron
As: *see* Group 15
atom-atom interaction model, 14, 28, 32, 57, 63, 74, 96*f*, 122, 124*f*, 150*f*
   for carbon molecules, 78
   for ML_2, 74
   for ML_3, 76
atom-atom repulsion model, 14, 123, 152
atomic orbital: *see* relevant orbital entries
aufbau principle: *see* electron configuration
augment: *see* point group
β-sheet: *see* protein
Bailar twist, 68, 126
Ba: *see* Group 2
Berry pseudo-rotation, 61, 64
BH_3, 28
B_2H_6, 28
Bi: *see* Group 15
bond
   two-centre, 77, 80, 87*f*, 134
   three-centre, 85, 87*f*

   covalent, 67*f*
   directional, 144
   F-F, 97
   glycosidic, 166
   in clusters, 139 *f*
   ionic, 67*f*
   length, 96
   localised, 82
   M-M, 143*f*
   overlap of orbitals, 81
   π, 76, 77, 106, 117
   phosphodiester, 164
   planar zig-zag, 83
   σ, 105, 110
   strength, 10, 80*f*, 96
bonding
   boranes, 85*f*
   covalent, 71, 112, 140
   dative, 111
   d-orbitals in, 95
   ionic, 71
   metal-ligand, 110*f*
boranes, 85*f*, 142*f*
   bonding, 86*f*
   pentaboranes, 91*f*
Born interpretation, 42
Born-Oppenheimer approximation, 9
BrF_3, 97, 98
butane, 79
Ca: *see* Group 2
carbon based chemistry, 77
carbonyl, 83
catenated carbon systems, 78
CFSE: *see* crystal field stabilisation energy
CFT: *see* crystal field theory
C_3H_6, 8
character, 42*f*
character table, 42*f*
   C_{2v}, 43
   D_3, 47
   O_h, 51
   T_d, 49

chelate bite, 122, 123
chiral, 6, 39f, 168
chloroform, 4
classical symmetry selection rule
  procedure, 63f
  normal mode, 63f
  rearrangement of polyhedra, 65
ClF_3, 97, 98
closo: *see* polyhedron
cluster
  borane, 85f
  transition metal, 139f
C_N: *see* coordination number
CO, 21, 140f
$Co_2(CO)_8$, 143, 154
$Co_4(CO)_{12}$, 157
$[Co_6(CO)_{14}]^{2-}$, 144
$[Co_6(CO)_{14}]^{4-}$, 144, 145, 149, 152, 159
$[Co_6(CO)_{15}]^{2-}$, 152, 154
$[Co_6C(CO)_{13}]^{2-}$, 159
$[Co(ethylenediamine)_3]^{3+}$, 10
$[Co(NH_3)_4Cl_2]^+$, 5
cohesive energy, 147
concerted mechanism, 64f
conformer, 6
  anti, 159
  endo, 159
  exo, 159
  gauche, 74, 159
  syn, 159
  trans, 74, 159
coordination number, 23, 73, 95f, 107
  definition, 23
  determination, 25, 89f
  determination for transition metal
    complexes, 111, 120f
  of 2, 74, 108
  of 3, 76, 108
  of 4, 76, 108
  of 5, 109, 133
  of 6, 109
  of 7, 110
  of 8, 110
  of 9, 110
  of lanthanides and actinides, 134
  of metal carbonyls, 140
  survey of for transition metal
    complexes, 108f
crystal field stabilisation energy, 112, 115
  and coordination number, 119
  of hexaquo systems, 115

ML_6, 118
  no π interactions, 117
  σ and π interactions, 117
crystal field theory, 106f, 112
  inadequacies, 116
  orbital energy level diagram, 114
  zero energy, 112
CSSRP: *see* classical symmetry
  selection rule procedure
Cu(II)-Ag(II)-Au(II) geometries, 108
$[Cu(H_2O)_6]^{2+}$, 3
cyclohexane, 4, 79
cytosine, 164f
Δ: *see* crystal field stabilisation energy
deltahedral, 57, 86, 145
diatomics, 14f
  F_2, 97
  homonuclear, 71
  see molecular orbital theory
  potential energy, 9
  see valence bond theory
1,2-dichloroethane, 7
1,2-dichlorethene, 7
dispersion force, 29
disulfide bridge, *see* protein
DNA, 163
  A-form, 167f
  B-form, 165f
  bases, 164f
  double helix, 167f
  poly[(dA-dT)]$_2$, 169
  sugar, 164
  Z-form, 167f
electron affinity, 72
electron configuration, 12
  aufbau principle, 17, 113
electron count
  metal valence, 118f, 144f, 141
electron deficient, 101
electron repulsion
  and cluster geometries, 147
eighteen-electron rule, 24, 112f, 119, 140f, 151
  ML_6, 118f
  non-octahedral systems, 119
eight-electron rule, 24, 73, 77, 140
electronegativity, 72f, 78f, 84, 97f
electronic effects, 14, 106, 140f
energy
  changes, 9
  clusters, 142f
  cohesive, 147

Gibb's free, 8
internal, 8
minimum, 3, 8
partitioning, 142f
potential, *see* potential energy
enantiomer, 6, 64
entropy, 8, 163
ETA: *see* extended topological approach
ethane, 7
extended topological approach, 88f
$[Fe_2(CO)_8]^{2-}$, 143, 155
$Fe_2(CO)_9$, 143, 155
$Fe_3(CO)_{12}$, 154f
$[Fe_6C(CO)_{16}]^{2-}$, 159
fluxional, 109, 155f
fo: *see* frontier orbital
force field, 32
fragment
  clusters as sum of, 149f
  formalism, 22
  molecular, 81
  of polyhedron, 86f
frontier molecular orbital: *see* molecular orbital
frontier orbital: *see* orbital
Ga: *see* Group 13
Ge: *see* Group 15
generating set: *see* point group
geometry
  definition, 5, 142
glucose, 81
glycine, 4
Group 2 compounds, 103
Group 13 compounds, 102f
Group 14 compounds, 101f
Group 15 compounds, 101f
Group 16 compounds, 100f
Group 17 compounds, 97f
Group 18 compounds, 100f
group (symmetry): *see* point group
guanine, 164f
H_2: *see* specific entries
halogen compounds: *see* Group 17
hexaquo systems
  crystal field splitting, 115
  geometry of, 131f
high spin / low spin, 115f, 121, 122, 129
ho: *see* hybrid orbital
HCN, 74
HCO, 74

Hoechst 33258, 3
$H_2Os_3(CO)_{10}$, 146
$H_2Os_6(CO)_{18}$, 155
$[HOs_6(CO)_{18}]^-$, 154
Hund's rules, 54, 113
hybrid orital: *see* orbital
hybridisation, 22, 77, 96f
hydrogen peroxide, 6
hydrophobic, 173
In: *see* Group 13
interactions
  1-3, 81
  1-4, 83
interhalides: *see* Group 17
ionic model, 135
ionisation energy, 72
$Ir_4(CO)_{12}$, 158
$Ir_6(CO)_{16}$, 154
isolobal analogy, 149f
isomer, 6
  of clusters, 154
isomerism
  conformational, 6
  optical, 6
  structural, 8, 156
isomerisation: *see* rearrangement and relevant geometries
Jahn-Teller
  effect, 2, 106, 108, 120, 130f
  theorem, 130f, 173
Langmuir-Blodgett film, 79
lanthanides, 132f
  contraction, 133
LFT: *see* ligand field theory
ligand
  electron donating properties, 139f
  encapsulated, 142, 145
  migration, 157
  π-donor, 116, 117, 131
  σ-donor, 116, 117f
  π-acceptor, 116, 117f, 129f, 131
ligand field theory, 106, 116f
ligand repulsion model, 122
linear dichoism, 167
Li_2O, 27
lone pair, 26
  stereochemically active, 102
  stereochemically inactive, 96, 102
low spin: *see* high spin / low spin
macromolecules, 163f
mannose, 81
$M_2(CO)_n$, 155

$M_3(CO)_n$, 155f
metal polyhedron: *see* polyhedron
methane, 2, 41
mfo: *see* molecular fragment orbital
Mg: *see* Group 2
ML_2, 74f, *103*f
ML_3, 76f, 97, 102f
ML_4, 76f, 98, 101f
ML_5, 97, 101
ML_6, 100
ML_7, 97
M_4L_n, 157
M_6L_n, 158
M_m, 143f
MM: *see* molecular mechanics
$[MnCl_6]^{4-}$, 10, 53f, 116
$Mn_2(CO)_{10}$, 154
mo: *see* molecular orbital
MO energy levels: *see* molecular orbital energy levels
MO theory: *see* molecular orbital theory
molecular fragment orbital: *see* molecular orbital
molecular geometry: *see* geometry
molecular orbital
  anti-bonding, 45f
  as sum of *ao's*, 15, 146
  bonding, 45f
  combination of *sao's*, 42, 45
  definition, 15
  fragment, 81, 111, 116
  from symmetry, 41f
  frontier, 90, 149
  localised, 22, 150
molecular orbital energy levels
  boron hydride fragments, 89
  diborane, 91
  ethane, 91
  from symmetry, 45f
  H_2, 15
  H_3, 46f
  HCN, 74
  HCO, 74
  heteronuclear diatomics, 21, 74
  homonuclear diatomics, 17f
  hydrocarbon fragments, 90
  $[MnCl_6]^{4-}$, 53f
  SiF_4, 46, 48f
  triatomics, 74f
MO theory, 14f, 17f
  applied to clusters, 146
  boranes, 86, 88

fragment formalism 22, 74, 81, 86f, 111
H_2, 1f
molecular mechanics: 14, 30, 78, 175
molecular structure
  definition, 5, 134
N: *see* Group 15
Ni(II)-Pd(II)-Pt(II) geometries, 121
nickel cyanide, 5
nido, *see* polyhedron
nobel gas: *see* Group 18
nodes, 45, 146
non-bonded radii, 14, 25f, 30, 122
normal mode, 63
  and reactions, 63f
nucleic acid, *see* DNA
nucleotide, 163f
octahedral complexes, 109, 111f
  rearrangement, 62
octet: *see* eight
operation: *see* symmetry operation
orbital, 11
  angular functions, 12, 13
  anti-bonding, 24, 43f, 112
  bonding 24, 45f, 112
  d, 12, 15, 48, 51, 95, 100, 111f, 132, 140 141
  d and wavefunction degeneracy, 130
  diffuse, 96
  energy, 12
  f, 132
  frontier, 90
  hybrid, 24, 73, 89, 90
  hybrid, 24, 27
  hydrogenic, 14
  molecular fragment, 90f, 144f
  non-bonding, 24, 45f
  non-hydrogenic, 12
  overlap, 10, 45f, 127, 122
  p, 12f, 48, 108, 112, 116, 132
  radial distribution function: *see* entry
  s, 12, 15, 105, 126f, 141
  s/p energy gap, 17, 96
  σ, π, 18, 76, 116f, 140
  symmetry adapted: *see* sao
  valence, 13f, 106, 110, 111, 116, 132, 144
$Os_3(CO)_{12}$, 156f
$Os_6(CO)_{18}$, 145
$[Os_6(CO)_{18}]^{2-}$, 145, 154
P: *see* Group 15
p block
  middle, 100

Index

π-acceptor: *see* ligand
π back-bonding, 140
π-donor: *see* ligand
parameters: *see* molecular modelling
Pauli exclusion principle, 14, 53
Pb: *see* Group 14
PE: *see* potential energy
penetration of electrons, 12, 73
pentaboranes, 90
peptide bond, *see* protein
periodic table, 11
    d-block, 11
    f-block, 11
    left hand side compounds, 102*f*
planar zig-zag, 83
point group, 37*f*
    achiral, 39, 180*f*
    chiral, 39, 179*f*
    generating sets, 39, 179*f*
    generation, 38, 179*f*
    Schoenflies notation, 38
polyhedral skeletal electron pair theory, 146
polyhedron
    arachno, 57, 86
    boranes, 86
    closo, 57, 86, 96, 84
    labelling notation, 57
    ligand, 57, 149*f*
    ligand determined by repulsion, 152*f*
    metal, 142*f*
    metal and ligand, 142, 150*f*
    ML_n, 58*f*, 107*f*
    nido, 57, 86
    rearrangement, 60*f*
    relaxation, 58*f*, 135
potential energy, 9, 31
    diatomic, 9
    molecular mechanics, 31
projection operator, 42*f*
protein, 163, 171*f*
    α-helix, 172*f*
    β-sheet, 172*f*
    β-turn, 172
    disulfide bridge, 173
    folding, 79, 174
purine, 164*f*
pyrimidine, 164*f*
radial distribution functions
    hybrid orbitals, 24
    hydrogen, 12
Ray-Dutt twist, 68, 126
reactant, 60*f*

reducible representation, 44*f*
$Rh_4(CO)_{12}$, 158
$Rh_6(CO)_{16}$, 145, 159
$[Rh_6C(CO)_{13}]^{2-}$, 151, 159
ring pucker, 80
$Ru_3(CO)_{12}$, 154, 156
S: *see* Group 16
sao: *see*
saw-horse, 98
Sb: *see* Group 15
$SbPh_5$, 27
Se: *see* Group 16
shielding of electrons, 12, 73, 133
Si: *see* Group 15
Sn: *see* Group 15
spectrochemical series
    ligand, 108, 115, 128
    metal, 115
spin pairing, 113
square planar complexes
    isomerisation, 68
Sr: *see* Group 2
SrF_2, 27
stable geometry, 10
stereochemical changes, 60*f*
stereoelectronic effects, 80*f*
stereoelectronics and reactivity, 84
steric effects, 1, 14, 108*f*, 120, 141*f*
steric model, 23*f*
steric plus electronic models, 14, 22, 57
steric vs. electronic, 106, 111, 120*f*, 140
structure
    definition, 5
    primary, 163*f*
    secondary, 163*f*
    tertiary, 163*f*
substitution reaction: *see* trans effect
sugar, 4, 6, 81*f*
    in DNA, 164, 171
    pyranose, 168
    see conformation
symmetry, 10*f*, 35*f*
    changes, 63
    cluster geometry, 143
    definition, 10
    of normal modes, 54
    of orbitals, 18
    of wavefunctions, 54
    operation: *see* point group
    relationship to energy, 10, 64
symmetry adapted orbital
    Cl_6, 53*f*
    F_4, 50*f*

H_3, 47
Mn, 54
$[MnCl_6]^{4-}$, 54
π-system of H_2CCHCH_2, 44
SiF_4, 50
symmetry operation
　definition, 36
　identity, 36
　improper, 36, 38, 18f
　inversion, 36, 180
　multiplication of, 37, 177f
　point, 36f, 177f, 179f
　see projection operator
　proper, 36, 38, 179f
　reflection, 36, 181
T-shape, 96
TA: *see* topological approach
Te: *see* Group 16
template, 4, 5, 57f, 85
　high symmetry, 10, 65f, 107f, 121
　second row systems, 76f
　symmetry, 69
　transition metal complexes, 110, 121, 127, 133
tensor surface harmonic theory, 136f
tetragonal: *see* Jahn-Teller effect
valence bond, 14
　H_2, 15
van der Waals
　attraction, 28
　radii, 26
VB theory: *see* valence bond

vibrations, 7f, 54f
　see normal mode
trans influence, 127f
transition metals, 105f
transition metal clusters, 139f
transition state, 39, 60, 61 VSEPR theory, 26f, 29, 30, 74f, 96f
　for carbon molecules, 78
　for ML_2, 75
　for ML_3, 76
W-effect, 83
Wade's rules
　for boranes, 86, 88f
　for transition metal clusters, 144f
Walsh Diagram, 75
*t*Tl: *see* Group 13
topological approach, 87f
thalidomide, 6
thymine, 164
torsion angle, 168, 174
trans effect, 127f
ris-chelate, 121
　geometry, 122f
　isomerisation, 62f, 67f, 125f
virial theorem, 18
wavefunction
　degeneracy, 123
　normalised, 44
　symmetry of, 56
zig-zag
　planar, 83
　Z-DNA, 168